# Online Stochastic Combinatorial Optimization

# Online Stochastic Combinatorial Optimization

Pascal Van Hentenryck and Russell Bent

The MIT Press
Cambridge, Massachusetts
London, England

This book was set in Computer Modern by the authors using LaTeX. Printed and bound in the United States of America.

Library of Congress Cataloging-in-Publication Data

Van Hentenryck, Pascal.
    Online stochastic combinatorial optimization   /   Pascal Van Hentenryck and Russell Bent.
        p.  cm.
    ISBN-13; 978-0-262-22080-4 (alk. paper)
    ISBN-10; 0-262-22080-6 (alk. paper)
    1. Stochastic processes.  2. Combinatorial optimization.  3. Online algorithms.
  4. Operations research.  I. Bent, Russell.  II. Title.

T57.32.V36  2006
003—dc22

                                                                                2006048141

10   9   8   7   6   5   4   3   2   1

# Contents

# Preface

*I don't have any solution, but I certainly admire the problem.*
— Ashleigh Brilliant

*I take the view, and always have done, that if you cannot say what you have to say in twenty minutes, you should go away and write a book about it.*
— Kord Brabazon

*Everything that happens once will never happen again. But everything that happens twice will surely happen a third time.*
— Paulo Coelho

**From A Priori to Online Optimization** Optimization systems traditionally have focused on a priori planning and are rarely robust to disruptions. The airline industry, a sophisticated user of optimization technology, solves complex fleet assignment, crew scheduling, and gate allocation problems as part of its operations using some of the most advanced optimization algorithms available. Yet unexpected events such as weather conditions or strikes produce major disruptions in its operations. In August 2005, British Airways took three to four days to resume its normal operations after a strike of one of its catering subcontractors; many of its planes and crews were at the wrong places because of the airline's inability to anticipate and quickly recover from the event. The steel industry, also a significant customer of optimization technology, typically separates strategic planning (which orders to accept) and tactical planning (which priorities to assign them) from daily scheduling decisions. The strategic decisions are based on coarse approximations of the factory capabilities and forecasts, sometimes leading to infeasible schedules, missed deadlines, and poor decisions when novel incoming orders arrive online.

Fortunately, the last decades have witnessed significant progress in optimization and information technology. The progress in speed and functionalities of optimization software has been simply amazing. Advances in telecommunications, such as the global positioning system (GPS), sensor and mobile networks, and radio frequency identification (RFID) tags, enable organizations to collect a wealth of data on their operations in real time.

It also is becoming increasingly clear that there are significant opportunities for optimization algorithms that make optimization decisions online. Companies such as UPS have their own meteorologists and political analysts to adapt their operations and schedules online. Pharmaceutical companies must schedule drug design projects with uncertainty on success, duration, and new developments. Companies such as Wal-Mart now try to integrate their supply chains with those of theirs suppliers, merging their logistic systems and replenishing inventories dynamically.

As a consequence, we may envision a new era in which optimization systems will not only allocate resources optimally: they will react and adapt to external events effectively under time constraints, anticipating the future and learning from the past to produce more robust and effective solutions. These systems may deal simultaneously with planning, scheduling, and control, complementing a priori optimization with integrated online decision making.

**Online Stochastic Combinatorial Optimization**  This book explores some of this vision, trying to understand its benefits and challenges and to develop new models, algorithms, and applications. It studies *online stochastic combinatorial optimization* (OSCO), a class of optimization applications where the uncertainty does not depend on the decision-making process. OSCO problems are ubiquitous in our society and arise in networking, manufacturing, transportation, distribution, and reservation systems. For instance, in courier service or air-taxi applications, customers make requests at various times and the decision-making process must determine which requests to serve and how under severe time constraints and limited resources.

Different communities approach new classes of problems in various ways. A problem-driven community studies individual applications and designs dedicated solutions for each of them. A theoretically oriented community often simplifies the applications to identify core algorithmic problems that hopefully are amenable to mathematical analysis and efficient solutions. These approaches are orthogonal and often produce invaluable insights into the nature of the problems. However, many professionals in optimization like to say that "there are too many applications with too many idiosyncratic constraints" and that "an approximated solution to a real problem is often preferable to an optimal solution to an approximated problem." As a result, this book takes a third, engineering-oriented, approach. It presents the design of abstract models and generic algorithms that are applicable to many applications, captures the intricacies of practical applications, and leverages existing results in deterministic optimization.

**Online Anticipatory Algorithms**  More precisely, to tackle OSCO applications, this book proposes the class of *online anticipatory algorithms* that combine online algorithms (from computer science) and stochastic programming (from operations research). Online anticipatory algorithms assume the availability of a distribution of future events or an approximation thereof. They take decisions during operations by solving deterministic optimization problems that represent possible realizations of the future. By exploiting insights into the problem structure, online anticipatory algorithms address the time-critical nature of decisions, which allows for only a few optimizations at decision time or between decisions.

The main purpose of this book is thus to present online anticipatory algorithms and to demonstrate their benefits on a variety of applications including online packet scheduling, reservation systems, vehicle dispatching, and vehicle routing. On each of these applications, online anticipatory algorithms are shown to improve customer service or reduce costs significantly compared to oblivious algorithms that ignore the future. The applications are diverse. For some of them, the underlying optimization problem is solvable in polynomial time. For others, even finding optimal solutions to the deterministic optimization where all the uncertainty is revealed is beyond the scope of current optimization software. Moreover, these applications capture different types of decisions. On some of them, the issue is to choose which request to serve, while, on others, the question is how to serve the request. On some of these applications, it is not clear even what the decisions should be in an online setting, highlighting some interesting modeling issues raised by online applications. In particular, the book presents some results on vehicle-routing strategies that were amazingly counterintuitive at first and seem natural retrospectively.

Of course, in practice, not all applications come with a predictive model of the future. The book also studies applications in which only the structure of model or historical data is available. It shows how to integrate machine learning and historical sampling into online anticipatory algorithms to address this difficulty.

**From Practice to Theory and Back**   Demonstrating the benefits of online anticipatory algorithms on a number of applications, however diverse and significant, is hardly satisfying. It would be desirable to identify the class of applications that are amenable to effective solutions by online anticipatory algorithms. Such characterizations are inherently difficult, however, even in deterministic optimization: indeed optimization experts sometimes disagree about the best way to approach a novel application. OSCO applications further exacerbate the issue by encompassing online and stochastic elements.

This book attempts to provide some insights about the behavior of online anticipatory algorithms by identifying assumptions under which they deliver near-optimal solutions with a polynomial number of optimizations. At the core of online anticipatory algorithms lies an anticipatory relaxation that removes the interleaving of decisions and observations. When the anticipatory relaxation is tight, a property called $\epsilon$-*anticipativity*, online anticipatory algorithms closely approximate optimal, a posteriori solutions. Although this is a strong requirement, the applications studied in this book are shown to be $\epsilon$-anticipative experimentally. The inherent temporal structure of these applications, together with well-behaved distributions, seems to account for this desirable behavior. The analysis presented here is only a small first step and much more research is necessary to comprehend the nature of OSCO applications and to design more advanced online anticipatory algorithms.

**A Model for Sequential Decision Making**   The theoretical analysis has an interesting side effect: it highlights the common abstract structure that was buried under the idiosyncracies of the applications, their models, and their algorithms. It also establishes clear links with *Markov Decision Processes* (MDP), a fundamental approach to sequential decision making extensively studied in artificial intelligence and operations research. Like MDPs, online stochastic combinatorial optimization alternates between decisions and observations, but with a subtle difference: the uncertainty in MDPs is endogenous and depends on the decider's actions. It is thus possible to define a variant of MDPs, called *Markov Chance-Decision Processes* (MCDPs), that captures online stochastic combinatorial optimization and whose uncertainty is exogenous. In MCDPs, the decision process alternates between observing uncertain inputs and deterministic actions. In contrast, the decision process in traditional MDPs, called *Markov Decision-Chance Processes* (MDCPs) here, alternates between actions whose effects are uncertain and observations of the action outcomes. As a consequence, MCDPs crystallize the essence of online stochastic combinatorial optimization that can be summarized as follows:

**Anticipatory Relaxation:** Contrary to MDCPs, MCDPs naturally encompass an anticipatory relaxation for estimating the future. The anticipatory relaxation can be approximated by the solutions of deterministic optimization problems representing scenarios of the future.

**Online Anticipatory Algorithms:** MCDPs naturally lead to a class of online anticipatory algorithms taking decisions online at each time step using the observations and approximations to the anticipatory relaxations.

**Anticipativity:** When the anticipatory relaxation is $\epsilon$-anticipative, online anticipatory algorithms produce near-optimal solutions with a polynomial number of optimizations.

**Learning:** The distribution of the inputs can be sampled and learned independently from the underlying decision process, providing a clean separation of concerns and computational benefits.


So, perhaps now that this book is written, twenty minutes are sufficient to describe its contents, making its writing yet another humbling experience. However, the open issues it raises are daunting both in theory and practice.

**Acknowledgments**   As Hervé Gallaire (and some other famous people) like to say, "we got by with a little help from our friends." Actually, more than a little help. First, we would like to express our gratitude to Eli Upfal and Luc Mercier who coauthored two chapters of this book. Eli is chair of the Department of Computer Science at Brown University and still finds time to talk to us. Luc Mercier has the annoying habit of asking all the difficult questions and the good habit of coming up with most of the answers. Both of them have improved our understanding of this topic tremendously. The research leading to this book started in the context of a National Science Foundation (NSF) ITR Award (DMI-0121495) in collaboration with Georgia Tech., MIT, and SUNY Buffalo. It is a pleasure to thank Shabbir Ahmed, Ismael De Farias, Michel Goemans, and George Nemhauser for all their insights and enthusiasm for research, and, of course, NSF for making it all possible. Several students and colleagues also contributed significantly to this research. Aris Anagnostopoulos is a master in debugging theoretical results and is the "man" in case you ever need — or want — to travel to Acapulco. Yannis Vergados started the research on online multiknapsacks and helped us finish it, although his passion is sport scheduling. Irit Katriel came up with the beautiful (but negative) result on suboptimality approximations. Laurent Michel has been a patient sounding board for the last five years. As always, many thanks to Bob Prior, our MIT Press editor, who has become an expert in the art of tightening and relaxing deadlines under uncertainty. Mel Goldsipe and her editors have been amazingly helpful, even correcting formulas! And, last but not least, to our families without whom this book would have remained only a possible realization of an uncertain future.

# 1 Introduction

*I learned many years ago never to waste time trying to convince my colleagues.*
— Albert Einstein

*If it's stupid but works, it isn't stupid.*
— *Amphibious Warfare Review*

This chapter provides the motivation underlying online stochastic combinatorial optimization and explains the synergy between a priori and online algorithms. It contrasts online stochastic optimization with online algorithms, stochastic programming, and Markov Decision Processes. The chapter also presents the structure and themes of this book.

## 1.1 From A Priori to Online Stochastic Optimization

Optimization systems traditionally have focused on a priori planning and scheduling. The airline industry solves complex optimization problems as part of its operations, using some of the most advanced optimization algorithms available, but it is only recently that airlines have started integrating robustness and uncertainty into their schedules [90, 91]. In August 2005, British Airways took several days to resume its normal operations after a strike of one of its catering subcontractors; many planes and crews were at the wrong places and the airline was unable to quickly recover from the unexpected event. The steel industry often separates strategic planning (which orders to accept), tactical planning (which priorities to assign them), and daily scheduling decisions. The strategic decisions use coarse approximations of the factory capabilities and forecasts, sometimes leading to infeasible schedules, missed deadlines, and poor decisions when orders arrive online. Moreover, emergency response systems are inherently online and operate in uncertain environments where resources must be allocated dynamically on a global scale. As pointed out by Eric Frost, codirector of San Diego State University's Immersive Visualization Center and a veteran of the tsunami relief effort: "Much of it is telecommunications, but it's really about how you use a whole bunch of things so that you are able to manage the resources for medicine, power, water and all the twenty or so major things that you need to do in the wake of a disaster."

Fortunately, the last decades have witnessed significant progress in optimization. They include significantly faster mathematical programming systems, the emergence of constraint programming as a fundamental tool for resource allocation, and a variety of online and approximation algorithms. Moreover, advances in telecommunications, such as GPS, sensor and mobile networks, and RFID tags, enable organizations to collect in real time a wealth of data that, with the right technology, can be used online to adaptively improve their operations.

It is thus becoming increasingly clear that there are significant opportunities for adaptive and integrated approaches to optimization. Recent research (for instance, [6, 11, 12, 13, 32, 35, 113]) has shown the benefits of adaptability for vehicle routing, packet scheduling, and resource allocation problems, exploiting stochastic information to produce better solutions. Companies such as UPS now employ meteorologists and political analysts to adapt their operations and schedules online. Pharmaceutical companies schedule drug design projects, incorporating uncertainty about success, duration, and new developments [31]. Recent research has also demonstrated the potential gains of integration. Integrated allocation of aircraft and flight times led to savings of about twenty million dollars a year [88] and companies such as Wal-Mart now try to integrate their supply chains with those of their suppliers, merging their logistic systems and replenishing inventories dynamically [63].

As a consequence, we envision a new era in which optimization systems will not only allocate resources optimally: they will react and adapt to external events effectively under time constraints, anticipating the future and learning from the past to produce more robust and effective solutions. These systems will deal simultaneously with planning, scheduling, and control, complementing a priori optimization with integrated online decision making. This new generation of optimization systems generates fundamental research issues, not only in combinatorial optimization, but also in many related areas. These systems will require novel integrations of traditional optimization technologies to quickly produce high-quality solutions. They will integrate optimization with machine learning and simulation to learn from the past, discover unexpected correlations, and anticipate the future. They will rely on statistical techniques to provide guaranteed performance under reasonable assumptions, while minimizing risk in life-threatening applications. And they will require radically different software systems, moving from offline modeling languages into optimization tools that will react to external events and return high-quality solutions under tight deadlines.

## 1.2   Online Stochastic Combinatorial Optimization

This book explores a small part of this vision, trying to understand its benefits and challenges and to develop new models, algorithms, and applications. It studies *online stochastic combinatorial optimization* (OSCO), a class of online optimization applications with exogenous uncertainty. OSCO applications are ubiquitous in our society and arise in networking, manufacturing, transportation, distribution, reservation systems, and emergency response systems. It is thus useful to review how online stochastic optimization algorithms may benefit our society.

**Ambulance Dispatching and Relocation**   Consider the problem of ambulance dispatching and diversion in conjunction with the emergency admission process. To develop a true emergency response system, ambulance dispatching and relocation must be considered globally for a number of hospitals in the same region [46]. The system should include

1.  all the interdependencies in the admission process where bottlenecks continuously evolve due to critical resources (for example, MRIs) whose operations have inherently uncertain durations;

2. the competition between elective and emergency patients for in-patient beds;

3. the complexity of transfer and discharge procedures.

Some of these issues (for example, ambulance dispatching) have been addressed by a priori optimization approaches [33] and, even though these approaches try to incorporate uncertainties (in general, by building redundancies), they fail to account for the inherently dynamic nature of emergency response systems. This book shows the significant benefits of online stochastic optimization for vehicle dispatching and reservation systems (see also [11, 12, 113]). Ambulance dispatching and diversion combine both problems in complex ways and there is hope that online optimization will be able to contribute to better solutions.

**Failures in the Power Grid**  Consider the problem of electric power generation and delivery over a complex power grid. Power stations are connected to various junctions, and power is generated by different types of generators at different costs. There are distinct startup and shutoff costs for different generators and significant delays between the time generators are activated and the time they actually deliver power to the grid. A generation and delivery plan must also satisfy the power limits of different cables. All these parameters are fixed and known in advance. The two major uncertainty factors are the varying power demands at different sites of the grid and unforeseeable power and delivery failures. These sources of uncertainty are different in nature and must be addressed accordingly. While power demand is a variable, it can be accurately modeled by a stochastic process with parameters learned from past data (as a function of time, date, and weather). One can thus obtain (a priori) a well-designed master plan requiring relatively negligible adjustments (in terms of extra costs) to correct for stochastic fluctuations. Faults, on the other hand, are rare events. Although there is considerable research in how to detect them quickly, it is too expensive to build a network and activate a plan that is immune to every possible (rare) fault. But failures on a power grid (as illustrated by the catastrophic multistate failure of August 14, 2003) propagate rapidly and must be contained to avoid shutting off the whole network or a significant portion of it. This is precisely where online optimization systems have a fundamental role to play: containing failures online under severe time constraints without exact knowledge on how the failures will propagate. Observe that the speed is a major factor in the quality of the decision and that past data and/or predictive models are available to predict how failures propagate.

**Pandemic Containment**  Consider the problem of containing pandemics using vaccination and quarantine. Once a pandemic is detected, which is a rare event, appropriate measures should be taken to contain it through combinations of vaccination and quarantine for various populations. This is again an online optimization problem where resources are limited (for example, the available doses of vaccine in an area are limited), constraints are imposed by government policies, and fast response is critical. It is inherently uncertain, since the rate of infection and the effectiveness of vaccines are highly unpredictable. Simplifications of the problem have been studied using a priori methodologies [72] and often assume that the infection rate and vaccine effectiveness are known. In fact the rate of spread of infection cannot be predicted precisely and there is a need for more adaptive techniques to determine — online and under strict time constraints and government policies — how best to allocate limited resources in order to contain major pandemics. These techniques will use predictive models on how pandemics evolve and how treatment may curb them, as well as distributions on how they can propagate geographically.

**Containment of Bushfires** Bushfires cause loss of life, destroy property, ruin the environment, and hurt tourism. Many efforts are under way to develop predictive models for the igniting and spreading of fires and the impact of various management and weather scenarios. Also under way is research on how to solve strategic planning problems in order to locate firefighter units and other resources optimally, which can be modeled as a two-stage stochastic integer program. However, when a bushfire actually happens, it becomes critical to deploy these limited resources dynamically and under uncertainty to contain the fire effectively.

## 1.3   Online Anticipatory Algorithms

The applications mentioned above demonstrate the potential of online stochastic combinatorial optimization and its synergy with a priori optimization under uncertainty. This book reports some first steps in building the technology to address them. The main innovation is to integrate stochastic programming (from operations research), online algorithms (from computer science), and combinatorial optimization for sequential decision making under uncertainty. The resulting paradigm has a number of desirable features:

1. Instead of searching for an a priori, optimal policy in extremely large search spaces, OSCO focuses on the current data and uses the uncertainty and optimization models to make decisions one at a time. There is a reasonable analogy with computer chess here. Chess (respectively OSCO) algorithms do not search for an a priori optimal policy for all games (respectively scenarios); they choose the next move based on the current position (respectively decisions) and a model of the adversary (respectively distribution).

2. OSCO may exploit complex uncertainty models including correlations between random variables. In fact, for some applications, OSCO does not even require a perfect uncertainty model. OSCO algorithms may learn, or approximate, the uncertainty model online using machine learning techniques and historical sampling. This is especially significant in applications such as pandemic containment, where the spread of infection and the effectiveness of vaccines are not known a priori, but can be learned as the pandemic develops.

3. OSCO naturally leverages progress in offline optimization to find high-quality and robust solutions quickly, and a priori methods to obtain robust, long-term strategic planning, thus boosting its performance and the quality of its decisions.

The main purpose of this book is to introduce a class of online anticipatory algorithms that make decisions online using samples of the future. Algorithm $\mathcal{E}$ is the most basic online anticipatory algorithm: it evaluates every possible decision at each time step, solving scenarios of the future with optimization technology to determine the best possible course of action. Unfortunately, algorithm $\mathcal{E}$ is computationally too demanding for many practical applications with time-critical decisions. Hence, this book explores two of its approximations, *consensus* ($\mathcal{C}$) and *regret* ($\mathcal{R}$), which provide similar benefits at a fraction of the cost.

The effectiveness of the online anticipatory algorithms is demonstrated in a variety of applications including online packet scheduling, reservation systems, vehicle dispatching, and vehicle routing. With each of these applications, online anticipatory algorithms are shown to improve customer service or reduce costs significantly compared to oblivious algorithms ignoring the future. The applications are conceptually and computationally diverse. In online packet scheduling, the goal is to select packets; in online reservations, it is to allocate resources to requests; in vehicle routing, the objective is to serve as many customers as possible while ensuring a feasible routing on multiple vehicles. The underlying optimization problems also vary greatly in complexity: they can be solved in polynomial time for packet scheduling, while they are extremely challenging for vehicle routing problems.

Demonstrating the value of online anticipatory algorithms is hardly sufficient and this book also attempts to justify their performance theoretically. This is not an easy enterprise, however, as is often the case in optimization. The key theoretical insight in this book is the $\epsilon$-anticipativity assumption that, informally speaking, implies that the arrival order of the inputs is not too significant. Under this assumption, the online anticipatory algorithms closely approximate the a posteriori optimal solution, that is, the optimal solution obtained if all the inputs had been revealed initially. This assumption largely holds for the applications in this book, probably due to their underlying temporal structure.

## 1.4   Online Stochastic Combinatorial Optimization in Context

The synergy between a priori and online optimization has been already highlighted in this chapter. But it is useful to contrast online stochastic combinatorial optimization with prior work in online algorithms, stochastic programming, and Markov decision processes in order to crystallize its potential contributions. The purpose is not to be comprehensive. Rather it is to position online stochastic combinatorial optimization in respect to some of the most relevant work and to convey the intuition behind the goals of this book.

### 1.4.1 Online Algorithms

Most of the work on online algorithms has been oblivious to the input distribution $\mathcal{I}$. The online algorithms use only the revealed inputs and past decisions to take the next decision. In general, competitive analysis [22, 36, 57] is used to analyze their performance of the resulting online algorithms, which means that the online algorithm is compared to an offline algorithm for the same problem. Consider a maximization problem $\mathcal{P}$ with an objective function $w$ and assume that $\mathcal{O}$ is an offline algorithm returning an optimal solution $\mathcal{O}(I)$ for $\mathcal{P}$ on a fully revealed input sequence $I$. The performance of an online algorithm $\mathcal{A}$ is measured by the ratio between the optimal offline value $w(\mathcal{O}(I))$ and its value $w(\mathcal{A}(I))$ over each possible input sequence $I$, that is,

$$\max_I \frac{w(\mathcal{O}(I))}{w(\mathcal{A}(I))}.$$

An online algorithm $\mathcal{A}$ is $c$-competitive if, for all sequences $I$,

$$w(\mathcal{O}(I)) \leq c\, \mathcal{A}(w(I)),$$

and $c$ is called the competitive ratio of algorithm $\mathcal{A}$.[1] Competitive analysis implicitly assumes a worst-case adversary who can choose the input sequence to induce the worst behavior of $\mathcal{A}$. The performance of $\mathcal{A}$ on these *pathological* inputs typically is not representative and hence competitive analysis may give only limited insights on its actual quality [66]. To address this limitation, online algorithms often are analyzed under some stochastic assumptions on the stream of inputs. For instance, the stochastic process generating the stream of requests might be stationary, periodic, or even bursty. See, for instance, [1, 23, 77] for some results along this direction.

It is only recently that researchers have begun to study how information about the input distribution may improve the performance of online algorithms. This includes scheduling problems [28, 13, 11, 14, 81, 104], online reservation [6, 113], vehicle routing problems [12, 25], elevator dispatching [82], bin packing [32], and inventory control [73]. For instance, list-scheduling algorithms exploiting the expected job durations were proposed in [81, 104] for online stochastic scheduling problems where the durations of jobs are independent random variables. The analysis does not compare the resulting solution to the best offline solution, but rather to the best list-scheduling algorithms [80]. The work of Goemans et al. [35, 45] on stochastic knapsack and covering problems is particularly interesting as it aims at quantifying the benefits of adaptability and offers some beautiful insights into the nature of uncertainty. The Sum-of-Squares algorithm [32] is an elegant algorithm for bin packing that recognizes the distribution online to avoid suboptimal behavior. Most of these results, however, are targeted at specific applications. Levi et al. [73] provide an elegant approximation algorithm with a worst-case performance of 2 for a stochastic inventory problem in which the demands are correlated temporally.

---

[1]Chapter 4 presents a 2-competitive algorithm for packet scheduling illustrating these concepts.

Online stochastic combinatorial optimization pushes this direction even further. OSCO algorithms have at their disposal a black-box to sample scenarios of the future and they exploit past and future information to take their decisions. The goal in OSCO is to maximize the expected profit of the online algorithm $\mathcal{A}$, that is,

$$\mathop{\mathbb{E}}_{\xi \in \mathcal{I}} w(\mathcal{A}(\xi)).$$

Moreover, the focus in OSCO is to develop abstract models and generic algorithms applicable to a variety of applications and suitable for the complex problems typically encountered in practice. This is why the OSCO algorithms rely on another black-box: the offline optimization algorithm $\mathcal{O}$.

### 1.4.2 Stochastic Programming

Decision making under uncertainty traditionally has focused on a priori optimization which, as mentioned earlier, is orthogonal and complementary to online optimization. This is the case with stochastic programming, which is more concerned with strategic planning than operational decisions at the core of online algorithms.

**Two-Stage Stochastic Linear Programs**  In two-stage stochastic programming, the decision variables are partitioned into two sets. The first-stage variables are decided before the realization of the uncertain parameters, while the second-stage or *recourse* variables are determined once the stochastic parameters are revealed. The name *recourse* comes from the fact that the decision for the first-stage variables may not be feasible and the second-stage variables make it possible to restore feasibility. Two-stage stochastic programming can thus be summarized as the decision process

Stage 1 decision $\rightarrow$ observe uncertainty $\rightarrow$ stage 2 decisions,
(now)                                                  (recourse)

and the goal is to compute a robust first-stage decision that minimizes the total expected cost. To illustrate two-stage stochastic programming, consider the following problem in which a company must decide how many units $x$ of a product it must produce under uncertainty about the demand of its customers [5]. It costs the company two dollars to produce product $P$ and there are two scenarios for the demand. In the first scenario, which has probability 0.6, the demand is 500. In the second scenario, the demand is 700. The company also has a recourse in case it does not produce sufficiently many units: it can buy product $P$ from a third party for 3 dollars. Since the number of scenarios is so small, one can write a linear program for this problem:

$$\begin{aligned}
\min \quad & x + (0.6 \times 3)y_1 + (0.4 \times 3)y_2 \\
\text{subject to} \quad & \\
& x + y_1 \geq 500 \\
& x + y_2 \geq 700.
\end{aligned}$$

In the linear program where all variables are nonnegative, $x$ is the first-stage variable, while $y_i$ is the second-stage variable representing how much to buy in scenario $i$. In general, the number of scenarios may be large or infinite, in which case it is infeasible to write the linear program explicitly as in our simple example. The standard formulation of a two-stage stochastic program is thus

$$\begin{aligned}
\min_{x \in X} \quad & \left\{ cx + \mathbb{E}_{w \in \mathcal{I}}[Q(x, \omega)] \right\} \\
\text{subject to} \quad & Q(x, \omega) = \min_{y \in Y} \{ f(\omega)y \mid D(\omega)y = h(\omega) - T(\omega)x \},
\end{aligned} \qquad (1.4.1)$$

where $X \subseteq \Re^{n_1}$, $c \in \Re^{n_1}$, $Y \subseteq \Re^{n_2}$, and $\omega$ is a random variable from a probability space $(\Omega, \mathcal{F}, \mathcal{I})$ with $\Omega \subseteq \Re^k$, $f : \Omega \to \Re^{n_2}$, $h : \Omega \to \Re^{m_2}$, $D : \Omega \to \Re^{m_2 \times n_2}$, and $T : \Omega \to \Re^{m_2 \times n_1}$. Elegant theoretical and algorithmic results have emerged for two-stage stochastic linear programs [20, 56].

**Two-Stage Stochastic Integer Programs**   When the second-stage variables are integral, stochastic integer programs are particularly challenging. Indeed the function $Q(\cdot, \omega)$ become nonconvex and possibly discontinuous [21, 89, 94, 95, 97, 106, 110]. This introduces severe difficulties in designing algorithms for stochastic integer programs. Even though the problem reduces to a specially structured integer program (IP) when the probability distribution $\mathcal{I}$ is finite, traditional integer programming techniques are not suited to exploit the special problem structure and deal with the problem dimensionality (see [96, 107, 110]). This area is the topic of intense research and directions include the lifting of valid inequalities from integer programs to stochastic integer programs [48, 47], as well as purely combinatorial approaches to specific stochastic integer programs [38].

**The Sample Average Approximation Method**   An alternative approach for stochastic programs with large sample spaces is the use of Monte-Carlo sampling to generate i.i.d. realizations $\omega^1, \ldots, \omega^N$, and approximate (1.4.1) with a sample average approximating (SAA) problem:

$$\min_{x \in X} \left\{ cx + \frac{1}{N} \sum_{i=1}^{N} Q(x, \omega^i) \right\}. \qquad (1.4.2)$$

Repeated solutions of (1.4.2) for different sample sizes along with statistical tests can provide approximate solutions together with estimates of the true optimal value. The SAA theory is well understood for stochastic linear programming [64, 101], but it is a fundamental research topic for stochastic integer programs.[2] Once again, it is interesting to observe the nature of the deterministic problems involved in the sample average approximation method. The resulting combinatorial optimization problems now feature a global objective function (the expected value over all scenarios) subject to a large number of feasibility constraints. Recent research indicates that these are the ingredients typically found in hard combinatorial optimization problems.

---

[2]Observe that the SAA method is an exterior sampling method. There also exist interior sampling methods where the sampling is integrated inside the algorithm [52].

| $d_1$ | $d_2$ | $p_1$ | $p_2$ | $p$ |
|-------|-------|-------|-------|------|
| 500 | 600 | 0.6 | 0.3 | 0.18 |
| 500 | 700 | 0.6 | 0.7 | 0.42 |
| 700 | 900 | 0.4 | 0.2 | 0.08 |
| 700 | 800 | 0.4 | 0.8 | 0.32 |

**Table 1.1:** Instance Data for a Simple 3-Stage Stochastic Program.

**Multistage Stochastic Programming**  The above formulations can be extended to multistage stochastic programs. Reconsider the simple example presented earlier for three stages [5]. The decision process can be summarized as

Stage 1 decision $\rightarrow$ observation $\rightarrow$ stage 2 decision $\rightarrow$ Observation $\rightarrow$ stage 3 decision.

The decision at stage 1 is similar to before. In stage 2, there is a recourse variable to meet the demand and another decision to decide what to buy for stage 3. Stage 3 consists of a final recourse to meet the demand at that stage.

Table 1.1 depicts the instance data for four scenarios. The first two columns specify the demand in periods 2 and 3. The next two columns give the probabilities of the demands at the respective stages. The last column specifies the probabilities of the scenarios. The stochastic program uses two types of variables:

- $x_s^i$ denotes the number of units to produce in stage $i$ ($1 \leq i \leq 2$) of scenario $s$ ($1 \leq s \leq 4$);
- $y_s^i$ denotes the recourse in stage $i$ ($2 \leq i \leq 3$) of scenario $s$ ($1 \leq s \leq 4$).

The resulting three-stage stochastic linear program is depicted in figure 1.1. There are three sets of constraints in this program. The first set of constraints imposes the demand constraints in stage 2, while the third set imposes the demand constraints in stage 3. The last set of constraints are particularly noteworthy. They usually are called the *nonanticipativity* constraints, as they prevent the stochastic program from anticipating the future. For instance, the constraint

$$x_1^1 = x_2^1 = x_3^1 = x_4^1$$

makes sure that the first-stage decision is the same for all scenarios. The constraints

$$y_1^2 = y_2^2 \ , \ x_1^2 = x_2^2$$

impose that scenarios 1 and 2 have the same second-stage recourse and decision, since they are not distinguishable at this stage. Indeed the demand $d_1$ is 500 for both scenarios. The last constraint

$$\begin{aligned}
\min \quad & 0.18(2x_1^1 + 3y_1^2 + 2x_1^2 + 3y_1^3) \; + \; 0.42(2x_2^1 + 3y_2^2 + 2x_2^2 + 3y_2^3) \; + \\
& 0.08(2x_3^1 + 3y_3^2 + 2x_3^2 + 3y_3^3) \; + \; 0.08(2x_4^1 + 3y_4^2 + 2x_4^2 + 3y_4^3)
\end{aligned}$$

$$\text{subject to}$$

$$x_1^1 + y_1^2 \geq 500$$
$$x_2^1 + y_2^2 \geq 500$$
$$x_3^1 + y_3^2 \geq 700$$
$$x_4^1 + y_4^2 \geq 700$$

$$x_1^1 + y_1^2 - 500 + x_1^2 + y_1^3 \geq 600$$
$$x_2^1 + y_2^2 - 500 + x_2^2 + y_2^3 \geq 700$$
$$x_3^1 + y_3^2 - 700 + x_3^2 + y_3^3 \geq 900$$
$$x_4^1 + y_4^2 - 700 + x_4^2 + y_4^3 \geq 800$$

$$x_1^1 = x_2^1 = x_3^1 = x_4^1$$
$$y_1^2 = y_2^2 \;, \; x_1^2 = x_2^2$$
$$y_3^2 = y_3^3 \;, \; x_3^2 = x_3^2$$

**Figure 1.1:** A Three-Stage Stochastic Linear Program.

has the same purpose for the last two scenarios. The *anticipatory relaxation* obtained by removing the nonanticipativity constraint plays a fundamental role in this book. It is also at the core of the dual decomposition method from [27]. They proposed a branch and bound algorithm for multistage stochastic programs in which bounding is achieved using the Lagrangian dual for the anticipatory relaxation and branching is performed on nonanticipativity constraints.

Swamy and Shmoys [108] have proposed a generalization of the SAA method to a broad class of multistage stochastic linear programs using a black-box for the distribution. However, in general, two-stage methods do not translate easily into efficient solutions to the related online or multistage problems. Shapiro [100] has shown that the SAA method cannot be extended efficiently to multistage stochastic optimization problems: the number of required samples must grow exponentially with the number of iterations, which is typically large or infinite in online applications. Combinatorial solutions suffer from similar exponential explosion.

**Relationship with Online Stochastic Combinatorial Optimization**    It is useful to make a number of observations to contrast online stochastic combinatorial optimization with multistage stochastic programming.

- Stochastic programming is an a priori optimization method. Existing, theoretical and experimental, results [100, 74] indicate that it is unlikely to scale to the large horizons and time constraints for the online applications contemplated in this book.

- As noted in [45], the nature of uncertainty is often different in online algorithms and stochastic programming. In stochastic programming, the uncertainty is mostly concerned with the "data": the customer demand in the above example or the set to cover in [103]. In contrast, the uncertainty in online stochastic combinatorial optimization is concerned with the variables: which packet to serve in packet scheduling, which customers to serve in vehicle routing, and which sets to use in set-covering.

- Online stochastic combinatorial optimization applications do not include a recourse — rather each decision is taken to preserve feasibility. Moreover, many online applications also include irrevocable decisions and service guarantees.

### 1.4.3  Markov Decision Processes

Markov decision processes (MDP) [87, 16] is another fundamental model for sequential decision making that has been studied intensively in the artificial intelligence and operations research communities. A detailed comparison between online stochastic combinatorial optimization and Markov decision processes is given in chapter 13, so only a brief discussion is included here.

An MDP is a quadruple tuple $\langle S, A, T, R \rangle$, where $S$ is the set of states, $A$ as the set of actions, $T$ is the transition function, and $R$ is the reward function. If $a \in A$ and $s, s' \in S$, $T(s, a, s')$ denotes the probability of reaching state $s'$ from state $s$ with action $a$ and $R(s, a)$ is the reward for taking action $a$ in state $s$. MDPs can have a finite or infinite horizon, in which case the rewards typically are discounted over time. For the applications considered in this book, discounts make little sense: a late customer in courier services or in an online reservation should not be discounted. As a result, it is natural to restrict attention to a finite horizon $H = [0, h]$ and to partition the states into

$$S = \bigcup_{t=0}^{h} S_t$$

where $S_0 = \{s_0\}$ and $s_0$ is the initial state. The solution to an MDP is a policy

$$\pi : S \rightarrow A$$

that specifies the action $a \in A$ to take in every state $s \in S$. The expected reward of a policy $\pi$, denoted by $V_0^\pi(s_0)$, is defined inductively as follows:

$$
\begin{aligned}
V_h^\pi(s_h) &= 0; \\
V_t^\pi(s_t) &= R(s_t, \pi(s_t)) + \sum_{s_{t+1} \in S} T(s_t, \pi(s), s_{t+1}) V_{t+1}^\pi(s_{t+1}) \qquad (0 \le t < h).
\end{aligned}
$$

The optimal policy $\pi^*$ for the MDP can be defined inductively as follows:

$$\pi^*(s_t) = \arg\max_{a \in A} \left( R(s_t, a) + \sum_{s_{t+1} \in S} T(s_t, a, s_{t+1}) V_{t+1}^{\pi^*}(s_{t+1}) \right).$$

Most of the work on MDPs has focused on computing optimal policies and falls in the realm of a priori methods. Once such an optimal policy is available, it can then be applied online without further computation.

MDPs can be used to model OSCO applications but they raise a number of fundamental issues. First, observe that the uncertainty is exogenous in OSCO: it does not depend on the decisions. In contrast, it is endogenous in MDPs where the uncertainty depends on what action is taken at each stage. To encode an OSCO application into an MDP, it is thus necessary to in-source the exogenous uncertainty, using states that represent the decisions, the revealed inputs, and the state of the distribution. The resulting MDPs become gigantic as they represent the product of the search space of an optimization problem and the possible inputs. Significant research has been, and is, devoted to addressing these limitations by reducing the state space to consider [31, 59, 85, 34]. But researchers face significant challenges in handling the large horizons considered in this book (see, for instance, [28]), as well as applications where the uncertainty is not fully characterized a priori.

A second issue arising when MDPs encode OSCO applications is the merging of the underlying optimization problems and uncertainty models. This makes it difficult to use the underlying optimization problem (which is somehow lost in translation), to sample the distribution, and to learn the distribution independently from the optimization problem. Generative models were used in [59] for sampling MDPs, but they do not address these issues satisfactorily. Chapter 13 introduces a variant of MDPs, called Markov Chance-Decision Processes (MCDP), that addresses these limitations and captures naturally the class of online anticipatory algorithms proposed in this book.

### 1.4.4 Partially Observable Markov Decision Processes

In some applications, the outcome of an action may not be directly observable. Partially Observable Markov Decision Processes (POMDP) [55] were introduced to address this limitation. A POMDP augments an MDP with two parameters, $\Omega$ and $O$, where $\Omega$ is the set of observations and $O$ is the observation function. More precisely, $O(s, a, o)$ is the probability of seeing an observation $o \in \Omega$ given that action $a$ was taken and ended in state $s$. One of the key components of POMDPs is the notion of a belief state. A belief state maintains the probabilities of being in particular states in the MDP given past observations. These probabilities are updated using Bayes' rule after each transition. For online stochastic combinatorial optimization, POMDPs share the same limitations as Markov decision processes and amplify them. See, for instance, [76] for a discussion of the complexity issues. The notion of belief states is, however, fundamental in the results of chapter 11, where machine-learning techniques are used to learn distributions online.

### 1.4.5 Performance of Online Algorithms

It is also useful to review how this book analyzes the performance of online algorithms. Consider a problem $\mathcal{P}$ and an input distribution $\mathcal{I}$. Let $\mathcal{A}$ be an online algorithm for $\mathcal{P}$, $\mathcal{O}$ be an optimization algorithm for $\mathcal{P}$, and $\pi^*$ be an optimal policy for an MDP $\langle S \cup \{s_0\}, A, T, R \rangle$ encoding $\mathcal{P}$ and $\mathcal{I}$. The following inequality

$$\underset{\xi \in \mathcal{I}}{\mathbb{E}} \; w(\mathcal{A}(\xi)) \; \leq \; \mathbb{E} \; \pi^*(s_0) \; \leq \; \underset{\xi \in \mathcal{I}}{\mathbb{E}} \; w(\mathcal{O}(\xi)) \tag{1.4.3}$$

summarizes the relationship between the algorithms. It indicates that the expected performance of the online algorithm $\mathcal{A}$ is dominated by the expected performance of the optimal policy $\pi^*$. When studying the performance of an online algorithm, this book analyzes the expected loss

$$\underset{\xi \in \mathcal{I}}{\mathbb{E}} \; (w(\mathcal{O}(\xi)) - w(\mathcal{A}(\xi))).$$

In particular, this book shows that, under the $\epsilon$-anticipativity assumption, the loss is small. Together with (1.4.3), this means that the online algorithm $\mathcal{A}$ also has a small loss with respect to the optimal policy $\pi^*$.

### 1.4.6 On the Synergy between A Priori and Online Optimization

It is important to comment briefly on the relative strengths and weaknesses of a priori and online optimization. In an ideal world, a priori optimization would compute an optimal policy to an accurate model of the application and apply the policy online with little or no additional cost. However, as mentioned earlier, such models need to account for numerous rare events that induce computational requirements beyond the capabilities of optimization technology. If a priori optimization is the only option, practitioners then need to simplify the model, obtaining optimal or near-optimal solutions to an approximated problem. Online optimization proposes an alternative road. It recognizes that uncertainties arising during operations are better handled online, where online algorithms can focus on the instance data, react to external events, anticipate the future, and learn the uncertainty online. As a result, online optimization avoids the need to search for policies in gigantic search space. The price to pay is the need to optimize online, which seems reasonable for many applications given their underlying nature and the progress in optimization software. Moreover, the synergy of a priori optimization (to compute robust architectures) and online optimization (to use these architectures adaptively) has the potential to find high-quality solutions to the real problems, not their approximations.

## 1.5 Organization and Themes

This book is organized in five parts. Parts I through III discuss three application areas in increasing order of complexity: online scheduling, online reservations, and online routing. They present online anticipatory algorithms for each application area and study their performance experimentally. Part I contains a first version of the theoretical analysis to build the intuition behind the anticipativity assumption, Part II contains an application of intermediate difficulty, and Part III is really the experimental core of this book and it tackles complex problems in vehicle routing and dispatching. Part IV relaxes the fundamental assumption underlying online anticipatory algorithms — the availability of a sampling procedure for the input distribution. It shows that, on some applications, the distribution can be learned online or replaced by historical data. Part V takes a step back: it recognize s that all these application areas, despite the different nature of their decisions and requirements, share the same abstract structure and are also solved by the same abstract algorithms. To formalize this insight, Part V proposes the concept of Markov Chance-Decision Process (MCDP) and studies its relationship to the traditional Markov Decision-Chance Process (MDCP). The online anticipatory algorithms can then be reformulated in this abstract setting highlighting their main fundamental computational elements: the underlying optimization problem, the anticipatory relaxation, the $\epsilon$-anticipativity assumption, and the clean separation between the optimization and uncertainty models. Part V also discusses future directions thar are mainly concerned with how to proceed when $\epsilon$-anticipativity does not hold.

Although the main focus is on the design, implementation, and evaluation of anticipatory algorithms for online stochastic optimization, there are many other themes in this book. Since only a few of them are clearly apparent from the book structure, it is useful to preview these themes in this introduction.

**The Value of Stochastic Information**   One of the main themes is to demonstrate the value of stochastic information for online combinatorial optimization. This book presents experimental evidence of the value of stochastic information by comparing oblivious and stochastic algorithms for online optimization and measuring their performances with respect to offline, a posteriori solutions. This is the beauty of online combinatorial optimization applications: once the uncertainty is revealed, the instance can be solved optimally and it is possible to assess the quality of the decisions. Reducing this gap on online stochastic vehicle routing has been a fundamental drive behind this research, offering a clear measure of progress.

**Anticipativity**   A second theme in this book is to provide experimental evidence validating the theoretical assumptions in the analysis of online anticipatory algorithms. For each class of applications, this book reports a number of experimental results along these lines and tries to understand the underlying intuition.

**The Tradeoff between Sampling and Optimization**   As is typically the case for difficult problems, the design space for online anticipatory algorithms is large and many different tradeoffs may need to be explored. It is thus not surprising that the tradeoff between the number of scenarios "solved" by the online algorithm and the quality of the optimization algorithm is also a major theme. In particular, the experimental results indicate that it is greatly beneficial to use more sophisticated optimization algorithms on fewer scenarios rather than weaker algorithms on many more scenarios. Once again, the claim is not that this is universally true. Simply, on the online applications considered in this book (which vary considerably in terms of complexity), the results show that sophisticated optimization algorithms typically need very few scenarios to overcome weaker methods on many more scenarios.

**Robustness**   The last theme of the book is the sensitivity of online anticipatory algorithms to noise in the input distribution. For most applications, this book provides experimental results demonstrating the performances of online anticipatory algorithms when the input distribution is noisy. As mentioned earlier, the book also studies how to learn the input distribution online and how to exploit historical data when no predictive model is available.

# I ONLINE STOCHASTIC SCHEDULING

# 2 Online Stochastic Scheduling

*The best thing about the future is that it only comes one day at a time.*
— Abraham Lincoln

This chapter presents the main algorithms for online stochastic combinatorial optimization. It presents a generic online stochastic scheduling problem that abstracts away the problem-specific objective function and constraints. As a result, the online algorithms are described at a high-level of abstraction and their building blocks are identified clearly. Subsequent parts in this book will adapt the algorithms to other contexts but many of the salient features are captured already here.

Section 2.1 presents the generic offline scheduling problem and section 2.2 presents its generic online counterpart. Section 2.3 introduces the generic online algorithms $\mathcal{A}$ that can be instantiated to a variety of oblivious and stochastic algorithms. Section 2.4 discusses the properties of the generic online scheduling problem and algorithm and contrasts them with online algorithms and stochastic optimization. Section 2.5 instantiates algorithm $\mathcal{A}$ to two oblivious algorithms often used in experimental comparisons. Then, sections 2.6, 2.7, and 2.8 describe the main online anticipatory algorithms considered in this book. Section 2.9 describes how to adapt the framework when the decision must be immediate but there is time between optimizations. Section 2.10 reviews the difficulty of the suboptimality approximation problem at the core of the regret algorithm.

## 2.1 The Generic Offline Problem

The offline problem assumes a discrete model of time and considers a time horizon $H = [1, h]$. The input is a sequence

$$\langle R_1, \ldots, R_h \rangle$$

where each $R_i$ is a set of requests. Each request $r$ can be served at most once. A solution to the offline problem serves a request $r$ from $R_1 \cup \ldots \cup R_t$ at each time $t \in H$ and can be represented as a sequence

$$\langle r_1, \ldots, r_h \rangle.$$

The goal is to find a solution $\langle r_1, \ldots, r_h \rangle$ satisfying the problem-specific constraints $C$ and maximizing the problem-specific objective function $w$ that is left unspecified in the generic problem.

To formalize the offline problem more precisely, we assume that the problem-specific constraints are specified by a relation $C$ over sequences and that the objective function is specified by a function $w$ from sequences to natural numbers. In other words, if $S$ is a sequence, $C(S)$ holds if the sequence $S$ satisfies the problem-specific constraints and $w(S)$ represents the profit of the sequence. It also is useful to assume the existence of a request $\perp$ that can be served at each time step, imposes no constraints, and has no profit. With these conventions, the offline problem can be formalized as

$$
\begin{aligned}
\max \quad & w(\langle r_1, \ldots, r_h \rangle) \\
\text{such that} \quad & \\
& C(\langle r_1, \ldots, r_h \rangle); \\
& \forall i \in H : r_i \in \bigcup_{t=1}^{i} R_t \setminus \{r_1, \ldots, r_{i-1}\} \cup \{\bot\}.
\end{aligned}
$$

In the following, we assume the existence of an algorithm $\mathcal{O}$ to solve the offline problem and use $\mathcal{O}(\langle R_1, \ldots, R_h \rangle)$ to denote the optimal solution for the input sequence $\langle R_1, \ldots, R_h \rangle$.

## 2.2 The Online Problem

In the online problem, the set of requests $R_t$ $(t \in H)$ are not known a priori but are revealed online as the algorithm executes. At each time $t \in H$, the set $R_t$ is revealed. Hence, at step $t$, the algorithm has at its disposal the sequences $\langle R_1, \ldots, R_t \rangle$ and the requests $\langle s_1, \ldots, s_{t-1} \rangle$ served in earlier steps and its goal is to decide which request to serve. More precisely, the online algorithm must select a request $s_t$ in the set

$$
\bigcup_{t=1}^{i} R_t \setminus \{s_1, \ldots, s_{i-1}\} \cup \{\bot\}
$$

such that the sequence $\langle s_1, \ldots, s_t \rangle$ satisfies the problem-specific constraints

$$
C(\langle s_1, \ldots, s_t \rangle).
$$

In the online problem, the sequence $\langle R_1, \ldots, R_h \rangle$ of sets of requests is drawn from a distribution $\mathcal{I}$. In other words, $\langle R_1, \ldots, R_h \rangle$ can be seen as the realizations of random variables $\langle \xi_1, \ldots, \xi_h \rangle$ whose distribution is specified by $\mathcal{I}$. Note that the distribution of $\xi_t$ may depend on $\xi_1, \ldots, \xi_{t-1}$. For instance, in the packet scheduling application in chapter 4, the sequence of packets is specified by Markov models. The online algorithms have at their disposal a procedure to solve the offline problem and a procedure to sample the distribution $\mathcal{I}$. Practical applications often include severe time constraints on the decision time and/or on the time between decisions. To model this requirement, the online algorithms may use only the optimization procedure only $\mathcal{T}$ times at each decision step.

## 2.3 The Generic Online Algorithm

The algorithms for online scheduling are all instances of the generic schema depicted in figure 2.1; they differ only in how they implement function CHOOSEREQUEST in line 3. The online algorithm starts with an empty sequence of requests (line 1) and then considers each time step in the schedule horizon (line 2). At each time $t$, the algorithm selects the request $s_t$ to serve, given the earlier served

ONLINE ALGORITHM $\mathcal{A}(\langle R_1, \ldots, R_h \rangle)$
1   $S_0 \leftarrow \langle \rangle$;
2   **for** $t \in H$ **do**
3     $s_t \leftarrow$ CHOOSEREQUEST$(S_{t-1}, \langle R_1, \ldots, R_t \rangle)$;
4     $S_t \leftarrow S_{t-1} : s_t$;
5   **return** $S_h$;

**Figure 2.1:** The Generic Online Algorithm.

request $S_{t-1}$ and the set $R_1, \ldots, R_t$ that have been revealed so far. The selected request $s_t$ is then concatenated with $S_{t-1}$ to produce the sequence $S_t$ of requests served up to time $t$. Observe that, on line 3, the call to CHOOSEREQUEST has access only to the set $R_1, \ldots, R_t$. The goal of the online algorithm is to maximize the expected profit

$$\mathbb{E}_{\xi_1, \ldots, \xi_h} \left[ \mathcal{A}(\langle \xi_1, \ldots, \xi_h \rangle) \right].$$

For this purpose, it has two main black-boxes at its disposal:

1. An optimization algorithm $\mathcal{O}$ that, given a sequence of past decisions $\langle s_1, \ldots, s_{t-1} \rangle$ and a sequence $\langle R_1, \ldots, R_h \rangle$ of sets of requests, returns an optimal solution

$$\mathcal{O}(\langle s_1, \ldots, s_{t-1} \rangle, \langle R_1, \ldots, R_h \rangle)$$

to the offline problem consistent with the past decisions, that is,

$$\max \quad w(\langle r_1, \ldots, r_h \rangle)$$
$$\text{such that}$$
$$C(\langle r_1, \ldots, r_h \rangle);$$
$$\forall i \in H : r_i \in \bigcup_{t=1}^{i} R_t \setminus \{r_1, \ldots, r_{i-1}\} \cup \{\bot\};$$
$$\forall i \in 1..t-1 : r_i = s_i.$$

Observe that $\mathcal{O}(\langle R_1, \ldots, R_h \rangle) = \mathcal{O}(\langle \rangle, \langle R_1, \ldots, R_h \rangle)$.

2. A conditional sampling procedure SAMPLE that, given a time $t$ and a sampling horizon $f$, generates a sequence

$$\text{SAMPLE}(t, f) = \langle R_{t+1}, \ldots, R_f \rangle$$

of realizations for the random variables $\xi_{t+1}, \ldots, \xi_f$ obtained by sampling the distribution $\mathcal{I}$.

**Notations**   In this book, a sequence of $k$ elements $s_1, \ldots, s_k$ is denoted $\langle s_1, \ldots, s_k \rangle$. If $S_1$ and $S_2$ are sequences, $S_1 : S_2$ denotes the concatenation of $S_1$ and $S_2$. When $S$ is a sequence and $e$ is not, the notation $S : e$ abbreviates $S : \langle e \rangle$. Since sequences are so important in specifying online algorithms, this book uses bold notations to specify some frequent ones. For instance, $\boldsymbol{R_i}$ denotes the sequence $\langle R_1, \ldots, R_i \rangle$. An expression of form $\xi_{i..j}$ denotes the sequence $\langle \xi_i, \ldots, \xi_j \rangle$. Element $t$ of a sequence $S$ is denoted by $S_t$. Moreover, the set of available requests for service at time $t$ given past decision $S_{t-1}$ and requests $\boldsymbol{R_t}$ is defined as

$$\textsc{Available}(S_{t-1}, \boldsymbol{R_t}) = \bigcup_{i=1}^{t} R_i \setminus \{s_1, \ldots, s_{t-1}\} \cup \{\bot\},$$

while the set of feasible requests at time $t$ is defined as

$$\textsc{Feasible}(S_{t-1}, \boldsymbol{R_t}) = \{r \in \textsc{Available}(S_{t-1}, \boldsymbol{R_t}) \mid C(S_t : r)\}.$$

For conciseness, in mathematical formulas, we abbreviate $\textsc{Feasible}(S_{t-1}, \boldsymbol{R_t})$ by $\mathcal{F}(S_{t-1}, \boldsymbol{R_t})$ and $s_t \in \textsc{Feasible}(\boldsymbol{s_{t-1}}, \boldsymbol{R_t})$ by $C(\boldsymbol{s_t}, \boldsymbol{R_t})$.

## 2.4   Properties of Online Stochastic Scheduling

Before presenting instantiations of the algorithm, we review some features of online stochastic scheduling or, more generally, online stochastic combinatorial optimization (OSCO). Indeed OSCO borrows some fundamental properties from both online algorithms and stochastic programming.

- OSCO inherits its algorithmic structure from online algorithms: the input is an online sequence of requests (or sets of requests) and the uncertainty concerns only which requests come and when. Once a request $r$ is revealed, all its associated information becomes available, that is, how it may impact the problem-specific objective function $w$ and constraints $C$.

- OSCO inherits a fundamental assumption from stochastic programming: the uncertainty does not depend on past decisions.

OSCO also features some significant differences from online algorithms and stochastic programming.

- Contrary to online algorithms, OSCO algorithms sample the distribution $\mathcal{I}$ to generate scenarios of the future at each decision time $t$.

- Contrary to stochastic programming, OSCO moves away from a priori optimization and uses optimization technology to make decisions online as the algorithm executes.

There are also some interesting properties of OSCO that deserve to be highlighted.

- The decision at time $t$ is constrained by earlier decisions and impacts future decisions. Indeed past decisions constrain which requests can be served at time $t$ and subsequently through the problem-specific constraints and objective function. For instance, the problem-specific constraints may include time windows on when a request may be served, as well as sequencing or capacity constraints.

- There are no *recourse* variables in OSCO. This is due to two fundamental properties of the generic algorithm. First, as discussed, the only uncertainty is on which requests come and not on their underlying data in the constraints and objective function. Second, feasibility always can be tested online and only partial feasible solutions are considered.

- OSCO algorithms solve optimization problems similar to the offline problem. As a consequence, OSCO naturally leverages existing deterministic algorithms for the offline problems and does not require the design and implementation of new deterministic algorithms.

Online stochastic scheduling can be viewed as an approach to the multistage stochastic program

$$\underset{\xi_1}{\mathbb{E}} \underset{\substack{s_1 \\ C(s_1, \boldsymbol{\xi_1})}}{\max} \quad \ldots \quad \underset{\xi_h}{\mathbb{E}} \underset{\substack{s_h \\ C(s_h, \boldsymbol{\xi_h})}}{\max} \quad w(\boldsymbol{s_h}).$$

The goal here is not to solve this stochastic problem a priori, since multistage stochastic programs can be extremely hard to solve or approximate [100]. Instead online stochastic algorithms take decisions online, one at a time, after each realization $\xi_t$. As a result, they do not "precompute" all decisions for all possible scenarios and focus on the instance data being revealed during the execution. Online stochastic algorithms also avoid the necessity of including recourses in the model, since they serve only feasible requests. Finally, in the presence of noisy distributions, the online algorithms may adapt to the instance data and tune the parameters of the distribution as the execution proceeds. The price to pay for these benefits comes from the need to solve optimization problems online during the execution of the algorithm. As a consequence, the goal of the online algorithms is to maximize the information they collect from the samples within their time constraints.

## 2.5   Oblivious Algorithms

To illustrate the generic online algorithm, it is useful to present two oblivious instantiations of SELECTREQUEST, that is, two implementations that do not sample the distribution. These instantiations are also used in the experimental results for evaluation of the online stochastic algorithms.

**Greedy (G):**   When each request $r$ contributes a profit $w(r)$ to the objective function, the greedy algorithm serves the feasible request with the highest reward. It can be specified as

CHOOSEREQUEST-$\mathcal{E}(s_{t-1}, R_t)$
1   $F \leftarrow$ FEASIBLE$(s_{t-1}, R_t)$;
2   **for** $r \in F$ **do**
3       $f(r) \leftarrow 0$;
4   **for** $i \leftarrow 1 \ldots \mathcal{T}/|F|$ **do**
5       $A \leftarrow R_t :$ SAMPLE$(t, h)$;
6       **for** $r \in F$ **do**
7           $f(r) \leftarrow f(r) + w(\mathcal{O}(s_{t-1} : r, A))$;
8   **return** $argmax(r \in F) \, f(r)$;

**Figure 2.2:** The Expectation Algorithm for Online Stochastic Scheduling.

CHOOSEREQUEST-G$(s_{t-1}, R_t)$
1   $F \leftarrow$ FEASIBLE$(s_{t-1}, R_t)$;
2   **return** $argmax(r \in F) \, w(r)$;

**Local Optimization (LO):**   This algorithm chooses the next request to serve at time $t$ by finding the optimal solution on the known requests. It can be specified as

CHOOSEREQUEST-LO$(s_{t-1}, R_t)$
1   $\gamma^* \leftarrow \mathcal{O}(s_{t-1}, \langle R_1, \ldots, R_t, \emptyset, \ldots, \emptyset \rangle)$;
2   **return** $\gamma_t^*$;

In LO, line 1 computes the optimal sequence $\gamma^*$ by calling the optimization algorithm on the chosen requests $s_{t-1}$, the revealed sets $R_t$, and empty sets for the (yet–to–be–revealed) sets $R_{t+1}, \ldots, R_h$. Line 2 simply returns the request scheduled at time $t$ in $\gamma^*$.

## 2.6   The Expectation Algorithm

This section presents the first online anticipatory algorithm. Algorithm $\mathcal{E}$ instantiates the generic algorithm $\mathcal{A}$ in which CHOOSEREQUEST is implemented by function CHOOSEREQUEST-$\mathcal{E}$ depicted in figure 2.2. Informally speaking, algorithm $\mathcal{E}$ generates scenarios by sampling the distribution and evaluates each feasible request against each scenario. Line 1 computes the requests that can be served at time $t$. Lines 2 and 3 initialize the evaluation function $f(r)$ for each feasible request $r$. The algorithm then generates a number of samples for future requests (lines 4 and 5). For each such sample, it computes a scenario $A$ consisting of all available and sampled requests (line 5). The algorithm then considers each feasible request $r$ at time $t$ (line 6) and applies the optimization algorithm for the scenario $A$ and the concatenation $s_{t-1} : r$, indicating that request $r$ is scheduled at time $t$ (line 7). The evaluation of request $r$ is incremented in line 7 with the profit of the optimal solution to the scenario $A$. All feasible requests are evaluated against all scenarios and the algorithm then returns the request $r \in F$ with the highest evaluation (line 8).

It is important to highlight line 4 where algorithm $\mathcal{E}$ distributes the optimizations across all available requests. Since there are $|F|$ possible requests to serve at time $t$ and the available decision time allows for only $\mathcal{T}$ optimizations, algorithm $\mathcal{E}$ evaluates each request $r$ against $\mathcal{T}/|F|$ scenarios. As a consequence, $\mathcal{T}$ should be large enough to produce high-quality results. Otherwise the possible choices are evaluated on only a small number of samples, possibly leading to poor decisions. For this reason, algorithm $\mathcal{E}$ needs to be approximated when time constraints are more severe, that is, when $\mathcal{T}$ is small or $|F|$ is large.

**Relationship to Stochastic Programming**   It is also interesting to relate Algorithm $\mathcal{E}$ to stochastic programming techniques. At time $t$, $\boldsymbol{\xi}_t$ has been revealed and the request $s_t$ ideally should be the solution to the multistage stochastic program[1]

$$\max_{\substack{s_t \\ C(s_t, \boldsymbol{\xi}_t)}} \quad \mathbb{E}_{\xi_{t+1}} \quad \max_{\substack{s_{t+1} \\ C(s_{t+1}, \boldsymbol{\xi}_{t+1})}} \quad \cdots \quad \mathbb{E}_{\xi_h} \quad \max_{\substack{s_h \\ C(s_h, \boldsymbol{\xi}_h)}} \quad w(s_h). \tag{2.6.1}$$

Once again, this stochastic program is typically very challenging and may contain a very large number of stages. Algorithm $\mathcal{E}$ approximates (2.6.1) in two steps. First, it relaxes the nonanticipativity constraints to obtain the problem

$$\max_{\substack{s_t \\ C(s_t, \boldsymbol{\xi}_t)}} \quad \mathbb{E}_{\xi_{t+1}, \ldots, \xi_h} \quad \max_{\substack{s_{t+1} \\ C(s_{t+1}, \boldsymbol{\xi}_{t+1})}} \quad \cdots \quad \max_{\substack{s_h \\ C(s_h, \boldsymbol{\xi}_h)}} \quad w(s_h) \tag{2.6.2}$$

where all the remaining uncertainty is revealed at time $t$. The problem (2.6.2) is approximated by sampling the distribution to obtain $n = \mathcal{T}/|F|$ deterministic problems of the form

$$\max_{\substack{s_t \\ C(s_t, \boldsymbol{\xi}_t^k)}} \quad \cdots \quad \max_{\substack{s_h \\ C(s_h, \boldsymbol{\xi}_h^k)}} \quad w(s_h) \tag{2.6.3}$$

where $\boldsymbol{\xi}_{t+i}^k = \langle \xi_1, \ldots, \xi_t, \xi_{t+1}^k, \ldots, \xi_{t+i}^k \rangle$ and $\xi_{t+1}^k, \ldots, \xi_{t+i}^k$ are realizations of the random variables $\xi_{t+1}, \ldots, \xi_{t+i}$ given the realizations of $\xi_1, \ldots, \xi_t$ $(1 \leq k \leq n)$. Problems of the form (2.6.3) are exactly

---

[1]We abuse notations and use the same symbol $\xi_i$ $(1 \leq i \leq t)$ to denote the random variable and its realization.

those that can be solved by algorithm $\mathcal{O}$. As a consequence, problem (2.6.2) is approximated by

$$\max_{\substack{s_t \\ C(s_t, \xi_t)}} \quad w(\mathcal{O}(s_{t-1} : s_t, \xi_h^1)) + \ldots + w(\mathcal{O}(s_{t-1} : s_t, \xi_h^n)). \tag{2.6.4}$$

Algorithm $\mathcal{E}$ solves problem (2.6.4) using $\mathcal{T}$ calls to $\mathcal{O}$, that is, it calls $\mathcal{O}(s_{t-1} : r, \xi_h^k)$ for each $r \in F$ and each $k$ in $1..n$. Each decision step in algorithm $\mathcal{E}$ is thus related to the application of the SAA method [64, 101], also an exterior sampling method, to the anticipatory relaxation. The main difference is that algorithm $\mathcal{E}$ does not aggregate the scenarios into a single deterministic problem, but rather solves a deterministic problem for each feasible request at time $t$. This design is motivated by the inherently discrete and combinatorial nature of the applications considered in this book: aggregating the scenario may result in a much harder problem. It also makes it possible to reuse existing algorithms directly or with minor changes. Note also that algorithm $\mathcal{E}$ approximates the anticipatory relaxations online at each decision step.

## 2.7 The Consensus Algorithm

*A consensus means that everyone agrees to say collectively what no one believes individually.*
— Abba Eban

Algorithm $\mathcal{E}$ distributes the available optimizations $\mathcal{T}$ across all requests. When $\mathcal{T}$ is small (due to the time constraints), each request is evaluated only with respect to a small number of scenarios and the algorithm may not yield much information. This is precisely why online vehicle routing algorithms [12] cannot use algorithm $\mathcal{E}$, since the number of requests is large (about fifty to one hundred), the decision time is relatively short, and optimizations are computationally expensive.

This section presents the consensus algorithm specifically designed to address this limitation. Algorithm $\mathcal{C}$ was introduced in [13] as an abstraction of the sampling method used in online vehicle routing [12]. Its key idea is to solve each scenario once and thus to examine $\mathcal{T}$ scenarios instead of only $\mathcal{T}/|F|$. More precisely, instead of evaluating each possible request at time $t$ with respect to each scenario, algorithm $\mathcal{C}$ executes the optimization algorithm once per scenario. The request scheduled at time $t$ in the optimal solution is credited with the scenario profit and all other requests receive no credit. More precisely, algorithm $\mathcal{C}$ is an instantiation of the generic online algorithm $\mathcal{A}$ using function CHOOSEREQUEST-$\mathcal{C}$ depicted in figure 2.3.

The main structure is similar to algorithm $\mathcal{E}$. The first five lines are the same. Line 1 computes the requests that can be served at time $t$. Lines 2 and 3 initialize the evaluation function $f(r)$ for each feasible request $r$. The algorithm then generates a number of samples for future requests (lines 4 and 5). Line 6 is particularly noteworthy as it optimizes the scenario given past decisions $s_{t-1}$ and thus chooses a specific request for time $t$. Line 7 increments the evaluation of the request $\gamma_t^*$

CHOOSEREQUEST-$\mathcal{C}(s_{t-1}, R_t)$
1   $F \leftarrow$ FEASIBLE$(s_{t-1}, R_t)$;
2   **for** $r \in F$ **do**
3      $f(r) \leftarrow 0$;
4   **for** $i \leftarrow 1 \ldots \mathcal{T}$ **do**
5      $A \leftarrow R_t :$ SAMPLE$(t, h)$;
6      $\gamma^* \leftarrow \mathcal{O}(s_{t-1}, A)$;
7      $f(\gamma_t^*) \leftarrow f(\gamma_t^*) + 1$;
8   **return** $argmax(r \in F)\ f(r)$;

**Figure 2.3:** The Consensus Algorithm for Online Stochastic Scheduling.

scheduled at time $t$ in the optimal solution $\gamma^*$ returned by $\mathcal{O}(s_{t-1}, A)$. Line 8 simply returns the request with the largest evaluation.

The main appeal of algorithm $\mathcal{C}$ is its ability to avoid partitioning the available optimizations among the requests (see line 4), which is a significant advantage when the number of optimizations is small. More precisely, algorithm $\mathcal{C}$ considers $\mathcal{T}$ scenarios, each of which is optimized once. Each optimization chooses the requests to be scheduled at times $t, t+1, \ldots, h$ contrary to algorithm $\mathcal{E}$, which chooses the requests at times $t+1, \ldots, h$ given a particular choice for time $t$. Its main drawback is its *elitism*. Only the request scheduled at time $t$ is given one unit of credit for a given scenario, while other requests are simply ignored. It ignores the fact that several requests may be essentially similar with respect to a given scenario. Moreover, it does not recognize that a request may never be the best for any scenario, but may still be robust overall. The consensus algorithm $\mathcal{C}$ behaves well on many vehicle-routing applications because, on these applications, the objective function is to serve as many customers as possible. As a consequence, at a time step $t$, the difference between the optimal solution and a nonoptimal solution is rarely greater than 1 at each time $t$, although losses accumulate over time. Hence, in some sense, the consensus algorithm approximates algorithm $\mathcal{E}$.

## 2.8   The Regret Algorithm

*For the majority of us, the past is a regret, the future an experiment.*
— Mark Twain

For a given scenario $A$, the consensus algorithm computes the optimal solution

$$\gamma^* = \mathcal{O}(s_{t-1}, A).$$

CHOOSEREQUEST-$\mathcal{R}(s_{t-1}, R_t)$
1   $F \leftarrow$ FEASIBLE$(s_{t-1}, R_t)$;
2   **for** $r \in F$ **do**
3       $f(r) \leftarrow 0$;
4   **for** $i \leftarrow 1 \ldots \mathcal{T}$ **do**
5       $A \leftarrow R_t :$ SAMPLE$(t, h)$;
6       $\gamma^* \leftarrow \mathcal{O}(s_{t-1}, A)$;
7       $f(\gamma_t^*) \leftarrow f(\gamma_t^*) + w(\gamma^*)$;
8       **for** $r \in F \setminus \{\gamma_t^*\}$ **do**
9           $f(r) \leftarrow f(r) + w(\widetilde{\mathcal{O}}(s_{t-1} : r, A, \gamma^*))$;
10  **return** $argmax(r \in F) \, f(r)$;

**Figure 2.4:** The Regret Algorithm for Online Stochastic Scheduling.

The expectation algorithm computes, in addition to $\gamma^*$, the suboptimal solution

$$\gamma^r = \mathcal{O}(s_{t-1} : r, A)$$

for each feasible request $r$ at time $t$ ($r \neq \gamma_t^*$). The key insight in the regret algorithm $\mathcal{R}$ is the recognition that, in many applications, it is possible to compute a good approximation $\widetilde{\gamma^r}$ to the suboptimal solution $\gamma^r$ quickly once the optimal solution $\gamma^*$ is available. As a result, it is possible to combine the ability of algorithm $\mathcal{E}$ to evaluate all requests on all samples while considering roughly as many scenarios as algorithm $\mathcal{C}$. Approximating the suboptimal solution $\mathcal{O}(s_{t-1} : r, A)$ given the optimal solution $\mathcal{O}(s_{t-1}, A)$ is called the suboptimality approximation problem [7].

Algorithm $\mathcal{R}$ is an instantiation of the generic online algorithm with the implementation of procedure CHOOSEREQUEST in figure 2.4. It relies on an algorithm $\widetilde{\mathcal{O}}$ that, given an optimal solution $\gamma^* = \mathcal{O}(s_{t-1}, R_h)$ and a request $r \in$ FEASIBLE$(s_{t-1}, R_t)$, returns an approximation to $\mathcal{O}(s_{t-1} : r, R_h)$, that is,

$$w(\widetilde{\mathcal{O}}(s_{t-1} : r, R_h, \gamma^*)) \leq w(\mathcal{O}(s_{t-1} : r, R_h)).$$

The algorithm follows the exact same structure as algorithm $\mathcal{C}$ but adds two lines (lines 8 and 9) to approximate the suboptimal solutions and to increase the evaluations of their respective requests. As a result, the algorithm is able to perform $\mathcal{T}$ optimizations (line 4) and to approximate the evaluation of each feasible request on every scenario.

It is necessary to impose some efficiency requirements on algorithm $\widetilde{\mathcal{O}}$ and, ideally, some guarantees on the quality of its approximation. Intuitively, the performance requirements impose that the $|F|$ suboptimality calls to $\widetilde{\mathcal{O}}$ do not take more time than the optimization call and that they return constant-factor approximations to the suboptimal solutions. The first definition of chapter 2 captures the quality requirements.

**Definition 2.1 (Suboptimality Approximation)** Let $\mathcal{P}$ be a scheduling problem and $\mathcal{O}$ be an offline algorithm for $\mathcal{P}$. Algorithm $\widetilde{\mathcal{O}}$ is a suboptimality approximation for $\mathcal{P}$ if, for every pair $(s_{t-1}, R_h)$ and every request $r \in \textsc{Feasible}(s_{t-1}, R_t)$,

$$w(\mathcal{O}(s_{t-1} : r, R_h)) \leq \beta \, w(\widetilde{\mathcal{O}}(s_{t-1} : r, R_h, \mathcal{O}(s_{t-1}, R_h)))$$

for some constant $\beta \geq 1$.

The next definition introduces amortized suboptimality approximations and imposes that the running time to compute all suboptimality approximations be no greater asymptotically than the time to compute the optimal solution they receive as input.

**Definition 2.2 (Amortized Suboptimality Approximation)** Let $\mathcal{P}$ be a scheduling problem, $\mathcal{O}$ be an offline algorithm for $\mathcal{P}$ running in time $g$, and $\widetilde{\mathcal{O}}$ be a suboptimality approximation algorithm for $\mathcal{P}$ running in time $\widetilde{g}$. Algorithm $\widetilde{\mathcal{O}}$ is an amortized suboptimality approximation for $\mathcal{P}$ if, for every pair $(s_{t-1}, R_h)$,

$$\sum_{r \in \textsc{Feasible}(s_{t-1}, R_t)} \widetilde{g}(s_{t-1} : r, R_h) \text{ is } O(g(s_{t-1}, R_h)).$$

In practice, it is not really necessary to compute the suboptimality approximation; it is sufficient to compute the *regret* of the suboptimality approximation, that is,

$$w(\mathcal{O}(s_{t-1}, R_h)) - w(\widetilde{\mathcal{O}}(s_{t-1} : r, R_h, \mathcal{O}(s_{t-1}, R_h))).$$

Note that the regret approximates the local loss of a suboptimal request,

$$\textsc{LocalLoss}(s_{t-1} : r, R_h) = w(\mathcal{O}(s_{t-1}, R_h)) - w(\mathcal{O}(s_{t-1} : r, R_h))$$

which represents the cost of scheduling a suboptimal request $r$ at time $t$, assuming that all subsequent decisions are optimal.

**Definition 2.3 (Local Loss)** Let $\mathcal{P}$ be a scheduling problem and $\mathcal{O}$ be an offline algorithm for $\mathcal{P}$. The local loss of a request $r$ for a pair $(s_{t-1} : r, R_h)$ $(r \in \textsc{Feasible}(s_{t-1} : r, R_t))$ is

$$\textsc{LocalLoss}(s_{t-1} : r, R_h) = w(\mathcal{O}(s_{t-1}, R_h)) - w(\mathcal{O}(s_{t-1} : r, R_h))$$

CHOOSEREQUEST-$\mathcal{R}(s_{t-1}, \boldsymbol{R_t})$
1   $F \leftarrow$ FEASIBLE$(s_{t-1}, \boldsymbol{R_t})$;
2   **for** $r \in F$ **do**
3     $f(r) \leftarrow 0$;
4   **for** $i \leftarrow 1 \ldots \mathcal{T}$ **do**
5     $A \leftarrow \boldsymbol{R_t}$ : SAMPLE$(t, h)$;
6     $\gamma^* \leftarrow \mathcal{O}(s_{t-1}, A)$;
7     $f(\gamma_t^*) \leftarrow f(\gamma_t^*) + w(\gamma^*)$;
8     **for** $r \in F \setminus \{\gamma_t^*\}$ **do**
9       $f(r) \leftarrow f(r) + w(\gamma^*) -$ REGRET$(s_{t-1} : r, A, \gamma^*)$;
10   **return** $argmax(r \in F) \, f(r)$;

**Figure 2.5:** The Regret Algorithm for Online Stochastic Scheduling Revisited.

**Definition 2.4 (Regret)** Let $\mathcal{P}$ be a scheduling problem and $\mathcal{O}$ be an offline algorithm for $\mathcal{P}$. The regret of a request $r$ for a pair $(s_{t-1} : r, \boldsymbol{R_h})$ ($r \in$ FEASIBLE$(s_{t-1} : r, \boldsymbol{R_t})$) receives an optimal solution $\mathcal{O}(s_{t-1}, \boldsymbol{R_h})$ and returns an upper approximation to the local loss of $r$ with respect to $(s_{t-1} : r, \boldsymbol{R_h})$, that is,

$$\text{REGRET}(s_{t-1} : r, \boldsymbol{R_h}, \mathcal{O}(s_{t-1}, \boldsymbol{R_h})) \geq \text{LOCALLOSS}(s_{t-1} : r, \boldsymbol{R_h}).$$

Algorithm $\mathcal{R}$ can be formulated in terms of regrets as shown in figure 2.5. The efficiency and quality guarantees also can be expressed in terms of regrets.

## 2.9   Immediate Decision Making

In some applications, the time to make a decision is very short and does not allow for many optimizations. However, there is some time available between decisions to optimize a number of scenarios. This is the case for the vehicle dispatching and routing applications in chapters 9 and 10. The question then is how to adapt the generic online algorithm to such an environment.

The key idea is to precompute the solutions to a number of scenarios that can then be used, at decision time, to make an informed choice. More precisely, the algorithm now maintains a set of scenario solutions over time. Between decisions, the online algorithm generates and optimizes as many scenarios as possible, augmenting the pool of available solutions. At decision time, the algorithm uses this pool to select a request to serve. The set of solutions must then be updated to remove solutions that are incompatible with the selected request. The idea of precomputation is particularly attractive for consensus and regret since they optimize every scenario exactly once.

ONLINE ALGORITHM $\mathcal{A}(\langle R_1, \ldots, R_h \rangle)$
1   $S_0 \leftarrow \langle \rangle$;
2   $\Gamma \leftarrow$ GENERATESOLUTIONS$(S_0, \langle \rangle)$;
3   **for** $t \in H$ **do**
4       $s_t \leftarrow$ CHOOSEREQUEST$(S_{t-1}, \boldsymbol{R_t}, \Gamma)$;
5       $S_t \leftarrow S_{t-1} : s_t$;
6       $\Gamma \leftarrow \{ \gamma \in \Gamma \mid \gamma_t = s_t \}$;
7       $\Gamma \leftarrow \Gamma \cup$ GENERATESOLUTIONS$(S_t, \boldsymbol{R_t})$;
8   **return** $S_h$;

GENERATESOLUTIONS$(\boldsymbol{s_t}, \boldsymbol{R_t})$
1   $\Gamma \leftarrow \emptyset$;
2   **repeat**
3       $A \leftarrow \boldsymbol{R_t} :$ SAMPLE$(t, h)$;
4       $\Gamma \leftarrow \Gamma \cup \{\mathcal{O}(\boldsymbol{s_t}, A)\}$;
5   **until** time $t + 1$
6   **return** $\Gamma$;

**Figure 2.6:** The Generic Online Algorithm with Precomputation.

Figure 2.6 depicts the generalized online algorithm. The set of scenario solutions $\Gamma$ is initialized in line 2. The request is selected in line 4 by function CHOOSEREQUEST, which now receives the set $\Gamma$ of scenario solutions as input as well. Line 6 removes the infeasible solutions, that is, those scenarios that do not schedule request $s_t$ at time $t$. Line 7 increases the number of plans by generating new solutions using the decisions and inputs at step $t$.

The functions GENERATESOLUTIONS and CHOOSEREQUEST now separate the optimizations from the decision-making logic. Function GENERATESOLUTIONS repeatedly generates (line 3) and solves (line 4) scenarios until the next decision time. The decision code is what is left in the instantiations of function CHOOSEREQUEST. Figure 2.7 shows how to instantiate function CHOOSEREQUEST for the consensus algorithm $\mathcal{C}$. The implementation uses the precomputed solution to select the decisions.

## 2.10   The Suboptimality Approximation Problem

*The older I get, the more I believe that at the bottom of most deep mathematical problems there is a combinatorial problem.*
— I.M. Gel'fand

CHOOSEREQUEST-$\mathcal{C}(s_{t-1}, R_t, \Gamma)$
1   $F \leftarrow$ FEASIBLE$(s_{t-1}, R_t)$;
2   **for** $r \in F$ **do**
3     $f(r) \leftarrow 0$;
4   **for** $\gamma \in \Gamma$ **do**
5     $f(\gamma_t) \leftarrow f(\gamma_t) + 1$;
6   **return** $argmax(r \in F) \, f(r)$;

**Figure 2.7:** The Consensus Algorithm with Precomputation.

It is useful to discuss suboptimality optimization problems in more detail to understand the underlying complexity issues. The regret algorithm is motivated by the "claim" that, once an optimal solution is available, it is reasonably easy to find approximations to the suboptimality problems. Is this really the case?

The suboptimality problem was studied in a more general context in [7] where the suboptimization problems consist of changing the value of a variable. For some hard–to–approximate problems, the suboptimality approximation problem is trivial once the optimal solution is available. This is the case for graph coloring. No constant factor approximation for graph coloring likely exists (unless $P = NP$) [75], yet the suboptimality problem can be solved exactly in polynomial time.

PROPOSITION 2.1   The suboptimality approximation problem can be solved exactly in polynomial time for graph coloring.

PROOF:   Let $O$ be a graph-coloring problem with optimal solution $\sigma^*$ and let $x$ be a variable whose color is $c^*$ in $\sigma^*$. The suboptimality problem that consists of assigning a color $c$ to $x$ can be solved optimally by swapping the colors $c^*$ and $c$ in $\sigma^*$.     $\square$

One may wonder if the availability of the optimal solution makes every problem simple to approximate. Unfortunately, the answer is negative: there are problems for which suboptimality approximation is as hard as the problem itself. One such problem is maximum satisfiability (MAX-SAT): given a CNF formula $\phi$, find a truth assignment that satisfies the maximum number of clauses.

PROPOSITION 2.2   Suboptimality of MAX-SAT is as hard as MAX-SAT.

PROOF:   Assume that there exists a polynomial-time (exact or approximate) suboptimality algorithm $\widetilde{\mathcal{O}}$ for MAX-SAT. We can construct an algorithm $\mathcal{O}$ that solves MAX-SAT (exactly or approximately) as follows. Given a CNF formula $\phi = (C_1 \wedge \cdots \wedge C_k)$ where each $C_i$ is a clause, $\mathcal{O}$

constructs a formula $\phi = (C'_1 \wedge \cdots \wedge C'_k)$, where $C'_i = (C_i \vee x)$ $(1 \leq i \leq k)$ and $x$ is a brand new variable. Obviously, any truth assignment in which $x$ is true is an optimal solution. $\mathcal{O}$ now calls $\widetilde{\mathcal{O}}$ on the formula $\phi'$, variable $x$, and any such optimal assignment. Since $\mathcal{O}$ returns or approximates the optimal solution for the case in which $x$ is assigned *false*, $\mathcal{O}$ returns or approximates the optimal solution for the original formula $\phi$.

$\square$

## 2.11 Notes and Further Reading

A preliminary, more specific version of the generic online scheduling problem and algorithm was presented in [13]. The main motivation was to re-express ideas from online stochastic vehicle routing [12] in a more general setting and to see whether they were beneficial to other problems. The paper presents algorithm $\mathcal{C}$ to formalize some of the techniques in [12] and compares algorithms $\mathcal{E}$ and $\mathcal{C}$ on packet scheduling. Algorithm $\mathcal{E}$ was used in [28] to solve the packet scheduling problem. The regret algorithm was introduced in [11] to bridge the gap between algorithms $\mathcal{E}$ and $\mathcal{C}$. Its name comes from a traditional branching heuristic in combinatorial optimization that selects first the variable with the largest difference on the objective function between its best and second–best values in its domain. The suboptimality approximation problem is studied in detail in [7], which contains the results presented earlier, as well as a number of suboptimality approximation algorithms.

# 3 Theoretical Analysis

Elizer Upfal and Pascal Van Hentenryck

*Prediction is very difficult, especially of the future.*
— Niels Bohr

This chapter provides a theoretical analysis of the solution quality and the runtime performance of the online stochastic algorithms. The analysis depends on the quality of the relaxation and on the number of scenarios solved at each time step. Its main result is to show that, under some assumptions that seem reasonably natural for the class of applications considered in this book, a relatively small number of scenarios per iteration suffices for achieving a high-quality soluton in the expected sense. Section 3.1 specifies that the goal of analysis is to determine the expected loss of algorithm $\mathcal{E}$ with respect to the optimal offline solution. It also specifies the main assumption about the analysis: the distribution must be anticipative for the problem. Section 3.2 shows that, under the assumption, the expected loss of algorithm $\mathcal{E}$ is the sum of the local errors at each time step. Section 3.3 bounds the local error at each time step. The main results are then given in section 3.4, while section 3.5 discusses the theoretical assumptions used in the analysis.

The results in this chapter are specific to online stochastic scheduling. The goal is to convey an initial intuition behind the anticipativity assumption and its consequences. Chapter 13 weakens the assumption, generalizes the analysis to the abstract setting of MCDPs, and shows why the experimental results collected in this book explain the good behavior of online anticipatory algorithms.

## 3.1 Expected Loss

The goal of this analysis is to determine the expected loss of algorithm $\mathcal{E}$ for an online scheduling problem $\mathcal{P}$, that is

$$\mathop{\mathbb{E}}_{\boldsymbol{\xi_h}} \left[ w(\mathcal{O}(\boldsymbol{\xi_h})) - w(\mathcal{E}(\boldsymbol{\xi_h})) \right],$$

where $\mathcal{O}$ is an optimization algorithm for $\mathcal{P}$ and $\boldsymbol{\xi_h}$ is drawn from the distribution $\mathcal{I}$. In other words, the analysis compares the behavior of algorithm $\mathcal{E}$ with the optimal offline solution in the expected sense. A similar result will be derived for algorithm $\mathcal{R}$.

The analysis is carried out under the hypothesis that the distribution $\mathcal{I}$ is anticipative for problem $\mathcal{P}$. This is clearly a strong assumption but to a large extent it holds for the applications and distributions considered in this book. Its meaning is discussed in detail in section 3.5.

**Definition 3.1 (Anticipative Distribution)** Let $\mathcal{P}$ be an online scheduling problem, $\mathcal{O}$ be an optimization algorithm for $\mathcal{P}$, and $\mathcal{I}$ be a distribution of instances of $\mathcal{P}$. Distribution $\mathcal{I}$ is anticipative for $\mathcal{P}$ if

$$\mathop{\mathbb{E}}_{\xi_{t..h}} \left[ w(\mathcal{O}(\boldsymbol{s_{t-1}}, \boldsymbol{R_{t-1}} : \xi_{t..h})) \right] = \mathop{\mathbb{E}}_{\xi_t} \left[ \max_{s_t \in \mathcal{F}(\boldsymbol{s_{t-1}}, \boldsymbol{R_{t-1}}:\xi_t)} \mathop{\mathbb{E}}_{\xi_{t+1..h}} \left[ w(\mathcal{O}(\boldsymbol{s_t}, \boldsymbol{R_{t-1}} : \xi_{t..h})) \right] \right]$$

for all times $t \in H$, realizations $\boldsymbol{R_t}$ of $\boldsymbol{\xi_t}$, and requests $\boldsymbol{s_{t-1}}$ such that $C(\boldsymbol{s_{t-1}}, \boldsymbol{R_{t-1}})$.

Intuitively, the anticipativity assumption means that, at each step $t \in H$, there is a "natural" request to select in order to maximize the expected profit.

Algorithm $\mathcal{E}$ computes a sequence of decisions

$$s_1 = \text{CHOOSEREQUEST-}\mathcal{E}(\boldsymbol{s_0}, \boldsymbol{\xi_1})$$
$$s_2 = \text{CHOOSEREQUEST-}\mathcal{E}(\boldsymbol{s_1}, \boldsymbol{\xi_2})$$
$$\ldots$$
$$s_t = \text{CHOOSEREQUEST-}\mathcal{E}(\boldsymbol{s_{t-1}}, \boldsymbol{\xi_t})$$
$$\ldots$$

Observe, once again, that the choice of request $s_t$ makes use only of $s_1, \ldots, s_{t-1}$ and $\xi_1, \ldots, \xi_t$.

## 3.2  Local Errors

Ideally, at time $t$, algorithm $\mathcal{E}$ should select the request

$$o^* = \underset{\substack{o \\ C(\boldsymbol{s_{t-1}} : o, \boldsymbol{R_t})}}{\text{arg-max}} \ \underset{\xi_{t+1..h}}{\mathbb{E}} \left[ w(\mathcal{O}(\boldsymbol{s_{t-1}} : o, R_t : \xi_{t+1..h})) \right]$$

with expected profit

$$\underset{\xi_{t+1..h}}{\mathbb{E}} \left[ w(\mathcal{O}(\boldsymbol{s_{t-1}} : o, R_t : \xi_{t+1..h})) \right].$$

In practice, $\mathcal{E}$ has time for only $\mathcal{T}$ optimizations and it selects a request $r \in \text{FEASIBLE}(\boldsymbol{s_{t-1}} : o, \boldsymbol{R_t})$ with expected profit

$$\underset{\xi_{t+1..h}}{\mathbb{E}} \left[ w(\mathcal{O}(\boldsymbol{s_{t-1}} : r, R_t : \xi_{t+1..h})) \right].$$

The difference between these two quantities is called the *local error* at time $t$.

**Definition 3.2 (Local Error at Time $t$)** Let $\mathcal{P}$ be an online scheduling problem, $\mathcal{O}$ be an optimization algorithm for $\mathcal{P}$, and $\mathcal{I}$ be a distribution for $\mathcal{P}$. The local error of a request $r$ at time $t$ for $(\boldsymbol{s_t}, \boldsymbol{R_t})$, denoted by $\Delta(\boldsymbol{s_{t-1}}, \boldsymbol{R_t}, r)$, is defined as

$$\underset{\substack{o \\ C(\boldsymbol{s_{t-1}} : o, \boldsymbol{R_t})}}{\max} \ \underset{\xi_{t+1..h}}{\mathbb{E}} \ w(\mathcal{O}(\boldsymbol{s_{t-1}} : o, \boldsymbol{R_t} : \xi_{t+1..h})) - \underset{\xi_{t+1..h}}{\mathbb{E}} \ w(\mathcal{O}(\boldsymbol{s_{t-1}} : r, \boldsymbol{R_t} : \xi_{t+1..h})).$$

When the distribution $\mathcal{I}$ is anticipative for $\mathcal{P}$, the expected loss of algorithm $\mathcal{E}$ is the sum of the local errors in each time step.

LEMMA 3.1 Let $\mathcal{P}$ be an online scheduling problem, $\mathcal{O}$ be an optimization algorithm for $\mathcal{P}$, and $\mathcal{I}$ be a anticipative distribution for $\mathcal{P}$. Then,

$$\sum_{t=1}^{h} \mathop{\mathbb{E}}_{s_t,\xi_t} [\Delta(s_{t-1}, \xi_t, s_t)] = \mathop{\mathbb{E}}_{s_h,\xi_h} [w(\mathcal{O}(\xi_h)) - w(s_h)].$$

PROOF: First observe that

$$\mathop{\mathbb{E}}_{s_t,\xi_t} \Delta(s_{t-1}, \xi_t, s_t) = \mathop{\mathbb{E}}_{s_t,\xi_t} [\max_{\substack{o \\ C(s_{t-1}) : o, \xi_t}} \mathop{\mathbb{E}}_{\xi_{t+1..h}} w(\mathcal{O}(s_{t-1}) : o, \xi_h)) - \mathop{\mathbb{E}}_{\xi_{t+1..h}} w(\mathcal{O}(s_t), \xi_h))],$$

which is equal to

$$\mathop{\mathbb{E}}_{s_t,\xi_{t-1}} [\mathop{\mathbb{E}}_{\xi_t} [\max_{\substack{o \\ C(\mathcal{E}(s_{t-1}) : o, \xi_t)}} \mathop{\mathbb{E}}_{\xi_{t+1..h}} w(\mathcal{O}(s_{t-1}) : o, \xi_h))] - \mathop{\mathbb{E}}_{\xi_{t..h}} w(\mathcal{O}(s_t), \xi_h))].$$

Since $\mathcal{I}$ is anticipative for $\mathcal{P}$, it follows that

$$\mathop{\mathbb{E}}_{s_t\xi_t} [\Delta(s_{t-1}, \xi_t, s_t)] = \mathop{\mathbb{E}}_{s_t,\xi_{t-1}} [\mathop{\mathbb{E}}_{\xi_{t..h}} [w(\mathcal{O}(s_{t-1}), \xi_h)]] - \mathop{\mathbb{E}}_{\xi_{t..h}} [w(\mathcal{O}(s_t), \xi_h))]]$$

$$= \mathop{\mathbb{E}}_{s_{t-1},\xi_h} w(\mathcal{O}(s_{t-1}), \xi_h)) - \mathop{\mathbb{E}}_{s_t,\xi_h} w(\mathcal{O}(s_t), \xi_h)).$$

Now consider the sequence of local errors:

$$\mathop{\mathbb{E}}_{s_1,\xi_1} [\Delta(s_0, \xi_1, s_1)] = \mathop{\mathbb{E}}_{\xi_h} [w(\mathcal{O}(\xi_h))] - \mathop{\mathbb{E}}_{s_1,\xi_h} [w(\mathcal{O}(s_1), \xi_h))]$$

$$\mathop{\mathbb{E}}_{s_2,\xi_2} [\Delta(s_1, \xi_2, s_2)] = \mathop{\mathbb{E}}_{s_1,\xi_h} [w(\mathcal{O}(s_1), \xi_h))] - \mathop{\mathbb{E}}_{s_2,\xi_h} [w(\mathcal{O}(s_2), \xi_h))]$$

$$\ldots$$

$$\mathop{\mathbb{E}}_{s_t,\xi_t} [\Delta(s_{t-1}, \xi_t, s_t)] = \mathop{\mathbb{E}}_{s_{t-1},\xi_h} [w(\mathcal{O}(s_{t-1}), \xi_h))] - \mathop{\mathbb{E}}_{s_t,\xi_h} [w(\mathcal{O}(s_t), \xi_h))]$$

$$\ldots$$

$$\mathop{\mathbb{E}}_{s_h,\xi_h} [\Delta(s_{h-1}, \xi_h, s_h)] = \mathop{\mathbb{E}}_{s_{h-1},\xi_h} [w(\mathcal{O}(s_{h-1}), \xi_h))] - \mathop{\mathbb{E}}_{s_h,\xi_h} [w(\mathcal{O}(s_h), \xi_h))].$$

It follows that

$$\sum_{t=1}^{h} \mathop{\mathbb{E}}_{s_t,\xi_t} \Delta(\mathcal{E}(\xi_t), \xi_t) = \mathop{\mathbb{E}}_{\xi_h} w(\mathcal{O}(\xi_h)) - \mathop{\mathbb{E}}_{s_h,\xi_h} w(\mathcal{O}(s_h), \xi_h))$$

$$= \mathop{\mathbb{E}}_{\xi_h} w(\mathcal{O}(\xi_h)) - \mathop{\mathbb{E}}_{s_h,\xi_h} w(s_h) = \mathop{\mathbb{E}}_{s_h,\xi_h} [w(\mathcal{O}(\xi_h)) - w(s_h)]. \qquad \square$$

## 3.3 Bounding Local Errors

It remains to bound the local errors. Let $\bar{s}_t$ be the random variable denoting the choice of algorithm $\mathcal{E}$ at step $t$, that is,

$$\bar{s}_t = \mathcal{E}(s_{t-1}, R_t).$$

We show that

$$\underset{\bar{s}_t}{\mathbb{E}}\,\Delta(s_{t-1}, R_t, \bar{s}_t) \leq \sum_{r \in \mathcal{F}(s_{t-1}, R_t)} \Delta(s_{t-1}, R_t, r)\, e^{-m(\Delta(s_{t-1}, R_t, r)^2/2\sigma_{t,r}^2}$$

where $m$ is the number of samples taken at each step and $\sigma_{t,r}$ is the standard deviation on the local loss of request $r$ at time $t$ for the distribution $\mathcal{I}$.

LEMMA 3.2  Consider a time $t$ and a pair $(s_{t-1}, R_t)$, and let $\bar{s}_t$ denote the random variable denoting the choice of algorithm $\mathcal{E}$ at step $t$. The expected local loss at step $t$ of algorithm $\mathcal{E}$ satisfies

$$\underset{\bar{s}_t}{\mathbb{E}}\,\Delta(s_{t-1}, R_t, \bar{s}_t) \leq \sum_{r \in \mathcal{F}(s_{t-1}, R_t)} \Delta(s_{t-1}, R_t, r)\, e^{-m(\Delta(s_{t-1}, R_t, r)^2/2\sigma_{t,r}^2}$$

where $m$ denotes the number of samples taken at step $t$ and $\sigma_{t,r}$ is the standard deviation on the local loss of request $r$ at step $t$.

PROOF:  At step $t$, algorithm $\mathcal{E}$ generates $m$ scenarios $R^1_{t+1..h}, \ldots, R^m_{t+1..h}$ and computes, for each request $r \in \mathcal{F}(s_{t-1}, R_t)$, an approximation

$$f(r) = \frac{1}{m} \sum_{k=1}^{m} w(\mathcal{O}(s_{t-1} : r, R_t : R^k_{t+1..h}))$$

of the expectation

$$\underset{\xi_{t+1..h}}{\mathbb{E}}\, w(\mathcal{O}(s_{t-1} : r, R_t : \xi_{t+1..h})).$$

Algorithm $\mathcal{E}$ chooses request $\bar{s}_t$, that is,

$$\bar{s}_t = \underset{r \in \mathcal{F}(s_{t-1}, R_t)}{\text{arg-max}}\; f(r).$$

Consider the request $s_t^*$ defined as

$$s_t^* = \underset{r \in \mathcal{F}(s_{t-1}, R_t)}{\text{arg-max}}\; \underset{\xi_{t+1..h}}{\mathbb{E}}\, [w(\mathcal{O}(s_{t-1} : r, R_t : \xi_{t+1..h}))].$$

If algorithm $\mathcal{E}$ chooses a request $r$ over $s_t^*$, it follows that

$$f(r) \geq f(s_t^*).$$

By definition of the local error,

$$\Delta(s_{t-1}, R_t, r) = \mathop{\mathbb{E}}_{\xi_{t+1..h}} w(\mathcal{O}(s_{t-1} : s_t^*, R_t : \xi_{t+1..h})) - \mathop{\mathbb{E}}_{\xi_{t+1..h}} w(\mathcal{O}(s_{t-1} : r, R_t : \xi_{t+1..h})).$$

Define $L_{t,r}$ as

$$(f(r) - f(s_t^*)) - (\mathop{\mathbb{E}}_{\xi_{t+1..h}} [w(\mathcal{O}(s_{t-1} : r, R_t : \xi_{t+1..h}))] - \mathop{\mathbb{E}}_{\xi_{t+1..h}} [w(\mathcal{O}(s_{t-1} : s_t^*, R_t : \xi_{t+1..h}))]]).$$

Since $f(r) - f(s_t^*) \geq 0$, it follows that

$$\Delta(s_{t-1}, R_t, r) \leq L_{t,r}, \tag{3.3.1}$$

which is a necessary condition for the selection of $r$ by $\mathcal{E}$. Moreover, the difference $f(r) - f(s_t^*)$ is the average of $m$ independent, identically distributed, random variables, each with mean

$$\mathop{\mathbb{E}}_{\xi_{t+1..h}} [w(\mathcal{O}(s_{t-1} : r, R_t : \xi_{t+1..h}))] - \mathop{\mathbb{E}}_{\xi_{t+1..h}} [w(\mathcal{O}(s_{t-1} : s_t^*, R_t : \xi_{t+1..h}))]$$

and with a standard deviation denoted by $\sigma_{t,r}$. By the central limit theorem, we can argue that

$$\sqrt{m} L_{t,r} / \sigma_{t,r}$$

is a normal distribution $N(0,1)$. Since

$$\mathop{\mathbb{E}}_{\bar{s}_t} [\Delta(s_t, R_t, \bar{s}_t)] = \sum_{r \in \mathcal{F}(s_{t-1}, R_t)} \Delta(s_{t-1}, R_t, r) \Pr(\bar{s}_t = r),$$

by the necessary condition (3.3.1), it follows that

$$\Pr(r = \bar{s}_t) \leq \Pr(\Delta(s_{t-1}, R_t, r) \leq L_{t,r})$$

and, by applying a Chernoff bound for the standard normal random variable [92], that

$$\Pr(\Delta(s_{t-1}, R_t, r) \leq L_{t,r}) \leq e^{-m(\Delta(s_{t-1}, R_t, r)^2 / 2\sigma_{t,r}^2}.$$

Hence,

$$
\begin{aligned}
\mathop{\mathbb{E}}_{\bar{s}_t} [\Delta(s_{t-1}, R_t, \bar{s}_t)] &= \sum_{r \in \mathcal{F}(s_{t-1}, R_t)} \Delta(s_{t-1}, R_t, r) \Pr(\bar{s}_t = r) \\
&\leq \sum_{r \in \mathcal{F}(s_{t-1}, R_t)} \Delta(s_{t-1}, R_t, r) \; \Pr(\Delta(s_{t-1}, R_t, r)) \leq L_{i,r}) \\
&\leq \sum_{r \in \mathcal{F}(s_{t-1}, R_t)} \Delta(s_{t-1}, R_t, r) \; e^{-m(\Delta(s_{t-1}, R_t, r))^2 / 2\sigma_{t,r}^2}.
\end{aligned}
$$

$\square$

## 3.4   The Theoretical Results

We are now in position to present the main result of this chapter.

THEOREM 3.1   Let $\mathcal{P}$ be an online scheduling problem, $\mathcal{O}$ be an optimization algorithm for $\mathcal{P}$, and $\mathcal{I}$ be a anticipative distribution for $\mathcal{P}$. The expected loss of algorithm $\mathcal{E}$ satisfies

$$\mathop{\mathbb{E}}_{\boldsymbol{\xi_h}}\left[w(\mathcal{O}(\boldsymbol{\xi_h})) - w(\mathcal{E}(\boldsymbol{\xi_h}))\right] \leq \sum_{t=1}^{h} \max_{\substack{s_{t-1} \\ \boldsymbol{R_t}}} \sum_{r \in \mathcal{F}(s_{t-1}, \boldsymbol{R_t})} \Delta(s_{t-1}, \boldsymbol{R_t}, r) e^{-m(\Delta(s_{t-1}, \boldsymbol{R_t}, r))^2 / 2\sigma_{t,r}^2},$$

where $m$ denotes the number of samples taken at each step of algorithm $\mathcal{E}$ and $\sigma_{t,r}$ denotes the standard deviation of the local loss of request $r$ at step $t$ for distribution $\mathcal{I}$.

PROOF:   Direct consequence of Lemmas 3.1 and 3.2.                                      □

This result has some noteworthy consequences. In particular, assuming that the $\sigma_{t,r}$ are $O(1)$, the expected loss of algorithm $\mathcal{E}$ is $O(1)$ when the number $m$ of samples taken at each step is $\Omega(\ln(h|R|))$. This induces $\Omega(|R|\ln(h|R|))$ optimizations per step and $|R|$ denotes the total number of requests.

COROLLARY 3.1   Let $\mathcal{P}$ be an online scheduling problem, $\mathcal{O}$ be an optimization algorithm for $\mathcal{P}$, and $\mathcal{I}$ be a anticipative distribution for $\mathcal{P}$. Assume that the standard deviations of the local losses are $O(1)$ for distribution $\mathcal{I}$. Then, algorithm $\mathcal{E}$, using $\Omega(\ln(h|R|))$ samples per iteration, has an expected loss of $O(1)$ and performs $\Omega(|R|\ln(h|R|))$ optimizations per step.

Consider now the regret algorithm $\mathcal{R}$ and assume that $\mathcal{R}$ uses a suboptimality approximation with factor $\rho$. Under this assumption, algorithm $\mathcal{R}$ returns an expected $\rho(1 + o(1))$-approximation of the optimal solution using $\Omega(\ln(h|R|))$ optimizations per step. Indeed theorem 3.1 also holds when $\mathcal{O}$ is replaced by its approximation $\tilde{\mathcal{O}}$ in the definitions and lemmas. As a consequence,

$$\mathop{\mathbb{E}}_{\boldsymbol{\xi_h}}\left[w(\tilde{\mathcal{O}}(\boldsymbol{\xi_h})) - w(\mathcal{R}(\boldsymbol{\xi_h}))\right]$$

is $o(1)$ for $\Omega(\ln(h|R|))$ samples. Since

$$w(\mathcal{O}(\boldsymbol{\xi_h})) \leq \rho \, w(\tilde{\mathcal{O}}(\boldsymbol{\xi_h})),$$

it follows that

$$w(\mathcal{O}(\boldsymbol{\xi_h})) \leq \rho \, (1 + O(1)) \, w(\tilde{\mathcal{O}}(\boldsymbol{\xi_h})).$$

COROLLARY 3.2 Let $\mathcal{P}$ be an online scheduling problem, $\mathcal{O}$ be an optimization algorithm for $\mathcal{P}$, and $\mathcal{I}$ be a anticipative distribution for $\mathcal{P}$. Assume that algorithm $\mathcal{R}$ uses a suboptimality approximation with factor $\rho$ and that the standard deviations of the local losses are $O(1)$ for the distribution $\mathcal{I}$. Then algorithm $\mathcal{R}$, using $\Omega(\ln(n|R|))$ samples per iteration, returns a $(\rho(1+O(1)))$-approximation of the optimal solution and performs $\Omega(\ln(n|R|))$ optimizations per step.

This result is significant in practice since it means that algorithm $\mathcal{R}$ approximates algorithm $\mathcal{E}$ while reducing the number of optimizations by a factor $|R|$. In general, it is not possible to obtain a similar result for consensus. However, we will come back to this issue in the context of vehicle routing applications, which have special structures.

## 3.5 Discussion on the Theoretical Assumptions

*The future arrives too soon and in the wrong order.*
— Alvin Toffler

This section briefly discusses the assumptions in the theoretical analysis on the anticipativity of the distribution and the standard deviations of the local losses.

### 3.5.1 The Meaning of Anticipativity

Consider an online scheduling problem $\mathcal{P}$ and an optimization algorithm $\mathcal{O}$ for $\mathcal{P}$. Recall that $\mathcal{I}$ is a anticipative distribution for $\mathcal{P}$ if

$$\underset{\xi_{t..h}}{\mathbb{E}}\left[w(\mathcal{O}(\boldsymbol{s_{t-1}}, \boldsymbol{R_{t-1}} : \xi_{t..h}))\right] = \underset{\xi_t}{\mathbb{E}}\left[\max_{\boldsymbol{s_t} \in \mathcal{F}(\boldsymbol{s_{t-1}}, \boldsymbol{R_{t-1}}:\xi_t)} \underset{\xi_{t+1..h}}{\mathbb{E}}\left[w(\mathcal{O}(\boldsymbol{s_t}, \boldsymbol{R_{t-1}} : \xi_{t..h}))\right]\right]$$

for all times $t \in H$, realizations $\boldsymbol{R_t}$ of $\boldsymbol{\xi_t}$, and requests $\boldsymbol{s_{t-1}}$ such that $C(\boldsymbol{s_{t-1}}, \boldsymbol{R_{t-1}})$. By induction on $t$, the anticipativity assumption amounts to requiring that, at each step $t$, the multistage stochastic program

$$\begin{array}{ccccccc} \max\limits_{\boldsymbol{s_t}} & \underset{\xi_{t+1}}{\mathbb{E}} & \max\limits_{\boldsymbol{s_{t+1}}} & \cdots & \underset{\xi_h}{\mathbb{E}} & \max\limits_{\boldsymbol{s_h}} & w(\boldsymbol{s_h}) \\ C(\boldsymbol{s_t}, \boldsymbol{R_t}) & & C(\boldsymbol{s_{t+1}}, \boldsymbol{R_t} : \xi_{t+1}) & & & C(\boldsymbol{s_h}, \boldsymbol{R_t} : \xi_{t+1..h}) \end{array}$$

is equivalent to its anticipatory relaxation

$$\begin{array}{ccccccc} \max\limits_{\boldsymbol{s_t}} & \underset{\xi_{t+1..h}}{\mathbb{E}} & \max\limits_{\boldsymbol{s_{t+1}}} & \cdots & & \max\limits_{\boldsymbol{s_h}} & w(\boldsymbol{s_h}). \\ C(\boldsymbol{s_t}, \boldsymbol{R_t}) & & C(\boldsymbol{s_{t+1}}, \boldsymbol{R_t} : \xi_{t+1}) & & & C(\boldsymbol{s_h}, \boldsymbol{R_t} : \xi_{t+1..h}) \end{array}$$

Is this a reasonable assumption in practice? At various places in this book, we will give some empirical evidence that this assumption is largely valid for the considered applications, instances, and distributions, most of which come from the literature. Here we provide some preliminary intuition to justify anticipativity on vehicle routing applications.

Consider an online vehicle problem with time windows. If a customer $c$ has a time window located between noon and 2:00 p.m., it is reasonable to assume that the request from $c$, if it occurs, will be earlier in the day, maybe as early as between 8:00 a.m. and 10:00 a.m. As a consequence, $c$ will be served long after its request time, since the vehicles are serving requests whose time windows are much earlier in the day. Generalizing this to many customers who are making requests for service later in the day, one can see that what really matters are which requests are coming and not the order in which they are coming, since they will be served much later in the day. As a consequence, anticipativity seems natural in this application.

Consider now the case where customer requests can be served immediately and must be serviced within a fixed time frame (they are lost otherwise). Once again, it does not seem beneficial in general to serve new requests immediately since they can be served later than the earlier requests. The exception of course is when a request can be accommodated easily now (for instance, it is close to the location of the vehicle) without disrupting the routing significantly. Once again, it seems that either the order of the requests is not important or there is a clear decision on the new request, making anticipativity rather natural.

There are, of course, applications where anticipativity does not hold. This happens, for instance, when the deadlines for serving the requests are very short and different. In such circumstances, the order of the next $k$ request may be significant. It is possible to generalize the assumption to $k$-anticipativity and to enhance the online algorithms to explore sequences of decisions, albeit at a higher computation cost. For instance, a distribution $\mathcal{I}$ is 2-anticipative for $\mathcal{P}$ if

$$\underset{\xi_{t..h}}{\mathbb{E}} \left[ w(\mathcal{O}(\boldsymbol{s_{t-1}}, \boldsymbol{R_{t-1}} : \xi_{t..h})) \right] = \underset{\xi_t}{\mathbb{E}} \Big[ \max_{s_t} \underset{\xi_{t+1}}{\mathbb{E}} \big[ \max_{s_{t+1}} \ldots \underset{\xi_{t+2..h}}{\mathbb{E}} \left[ w(\mathcal{O}(\boldsymbol{s_{t+1}}, \boldsymbol{R_{t-1}} : \xi_{t..h})) \right] \big] \Big]$$

where

$$\begin{aligned} s_t &\in \mathcal{F}(\boldsymbol{s_{t-1}}, \boldsymbol{R_{t-1}} : \xi_t) \\ s_{t+1} &\in \mathcal{F}(\boldsymbol{s_t}, \boldsymbol{R_{t-1}} : \xi_t : \xi_{t+1}) \end{aligned}$$

for all times $t \in H$, realizations $\boldsymbol{R_t}$ of $\boldsymbol{\xi_t}$, and requests $\boldsymbol{s_{t-1}}$ such that $C(\boldsymbol{s_{t-1}}, \boldsymbol{R_{t-1}})$. Chapter 13 discusses how to generalize algorithm $\mathcal{E}$ under these circumstances.

Finally, there are cases where the distribution is not $k$-anticipative for any $k$. Example 4.1 in chapter 4 illustrates such a distribution for packet scheduling. In such cases, the algorithm cannot select a request that is "good" for all realizations of the random variables. It is not an intrinsic limitation of the algorithm per se. Indeed all online algorithms will suffer from the same difficulty. The online algorithms presented in this book may in fact select the "best" request in the expected sense, but the resulting solutions may be far from the corresponding optimal offline solutions. Other kinds of theoretical analyses are necessary in these cases.

### 3.5.2 Standard Deviations on the Local Losses

Corollaries 3.1 and 3.2 also assume that the standard deviations of local losses for the distribution $\mathcal{I}$ are constant. Recall that the local loss at step $t$ is

$$w(\mathcal{O}(s_{t-1}, R_h)) - w(\mathcal{O}(s_{t-1} : r, R_h)).$$

Intuitively, this means that a single wrong decision cannot induce a significant degradation in quality and/or can be corrected in the future. Once again, this assumption is natural in vehicle routing applications where the goal is to serve as many customers as possible and for those applications, such as packet scheduling and reservation systems, where the differences in request rewards are bounded by a constant.

## 3.6 Notes and Further Reading

An earlier version of the results in this chapter is available in [112]. We must mention the work of Meyerson on online randomized oblivious algorithm for uncapacitated facility location [77]. Meyerson's analysis of the algorithm considers input sequences where the requests can be chosen by an adversary but not by the order in which they arrive. Although this is a different context, it is interesting to see the importance of ordering in the analysis.

# 4 Packet Scheduling

*When the going gets tough, the tough get empirical.*
— Jon Carroll

This chapter considers an online packet scheduling studied in [28, 60]. It shows how to apply the online anticipatory algorithms from chapter 2 to these problems, provides a comparison of their runtime performance and quality, and discusses the assumptions from chapter 3 in this context. Packet scheduling is particularly pertinent for two main reasons.

1. The number of requests to consider at each time $t$ is small and the offline problem can be solved in polynomial time. Hence, it is possible to evaluate and compare all the algorithms experimentally, which is not the case in the vehicle routing and dispatching applications where algorithm $\mathcal{E}$ is not practical. As a result, the experimental results gives us some insight into the loss incurred by using algorithms $\mathcal{C}$ and $\mathcal{R}$ instead of algorithm $\mathcal{E}$.

2. The uncertainty in this application is specified by a collection of Markov Models (MMs), one for each packet type [28]. It thus leads to an online stochastic application with dependencies between the random variables at different times in the horizon.

Sections 4.1 and 4.2 present the packet scheduling problem and the optimization algorithm. Section 4.3 studies the behavior of the greedy algorithm. Section 4.4 gives the suboptimality approximation for packet scheduling. Section 4.5 describes the experimental setting and section 4.6 presents the experimental results. Section 4.7 studies the anticipativity assumption for packet scheduling.

## 4.1 The Packet Scheduling Problem

**Relevance**  Future networks are expected to support Quality of Service (QoS) features in addition to the "best effort" service provided by the Internet today [60]. These QoS features must provide guarantees on packet loss, delays, and other relevant measures of performance. The packet scheduling application studied in this chapter is based on a proposal for integrating of QoS in the IP framework and it implements the so-called uniform bounded delay model.

**Informal Description**  The offline problem consists of scheduling packets over a finite horizon $H = [1, h]$. Its input is a sequence $\langle R_1, \ldots, R_h \rangle$, each $R_t$ being a set of packets. The packets are partitioned into a set of classes $C$ and each packet $j$ is characterized by its weight $w(j) \geq 0$, its arrival date $a(j) \in H$, and its class $c(j) \in C$. A packet $j \in R_t$ has an arrival date $a(j) = t$. Packets in the same class have the same weight but different arrival times. In other words, a packet is uniquely identified by a pair $\langle c, a \rangle$, where $c$ is a class and $a$ is an arrival time. Each packet $j$ requires a single time unit to process and must be scheduled in its time window $[a(j), a(j)+d]$, where $d$ is the same constant for all packets (that is, $d$ represents the time a packet remains available to schedule). At most one packet can be scheduled at each time $t$ and packets that cannot be served in their time windows are dropped. The goal is to find a schedule of maximal weight, that is, a schedule that maximizes the sum of the weights of all scheduled packets. This is equivalent to minimizing weighted packet loss.

**Formalization**   Packet scheduling is an instantiation of the generic scheduling problem presented in chapter 2 where the objective function is specified as

$$w(\langle s_1, \ldots, s_h \rangle) = \sum_{i=1}^{t} w(s_i),$$

and the problem-specific constraints $C(\langle r_1, \ldots, r_t \rangle)$ are specified as follows:

$$C(\langle r_1, \ldots, r_t \rangle) \equiv \forall i \in 1..t : a(r_i) \leq i \leq a(r_i) + d.$$

## 4.2   The Optimization Algorithm

This offline problem can be solved in time $O(|R||H|)$ [28]. This section reviews the optimization algorithm and proposes two useful refinements in an online context.

### 4.2.1   The Basic Optimization Algorithm

The algorithm, which is loosely based upon earlier work in [114], is depicted in figure 4.1. It considers the requests one at a time in the decreasing order

$$\langle w(r), a(r) \rangle.$$

In other words, the algorithm starts first with the request of highest weight and, in case of ties, with the request with latest arrival time (line 4). It tries to schedule each such request $r$ as late as possible before its deadline $a(r) + d$ (line 5). If such a time $p$ exists for packet $r$ (line 6) and is after its arrival date $a(r)$ (line 7), packet $r$ is included in the schedule (line 8). Otherwise, the algorithm tries to shuffle the current schedule to accommodate request $r$ (line 10).

The shuffling process tries to move an earlier request (whose weight is at least as high as request $r$) to time $p$. It searches for the leftmost time $q$ whose requests can be scheduled in time $p$ (line 4). If such a time exists, the shuffling procedure swaps the packets in time $p$ and time $q$ and reiterates the process until packet $r$ has been scheduled after its arrival date (line 3) or until no packet can be swapped with packet $r$ at some step (line 5). In this last case, the shuffling procedure returns the current schedule (line 9). Otherwise, the new schedule now includes request $r$ and has shuffled a number of packets leftward.

ALGORITHM $\mathcal{O}(\langle R_1, \ldots, R_h \rangle)$

  1   $R \leftarrow \{r \in R_t \mid 1 \le t \le h\}$;

  2   **for** $t \in H$ **do**

  3       $\gamma_t \leftarrow \bot$;

  4   **for** $r \in R$ ordered by decreasing $\langle w(r), a(r) \rangle$ **do**

  5       $p \leftarrow max\{t \in H \mid t \le a(r) + d \; \& \; \gamma_t = \bot\}$;

  6       **if** $p \ne -\infty$ **then**

  7         **if** $p \ge a(r)$ **then**

  8           $\gamma_p \leftarrow r$;

  9         **else**

10           $\gamma \leftarrow \text{SHUFFLE}(\gamma, r, p)$;

11   **return** $\gamma$;


SHUFFLE$(\gamma, r, p)$

  1   $\gamma' \leftarrow \gamma$;

  2   $\gamma_p \leftarrow r$;

  3   **while** $p < a(r)$ **do**

  4       $q \leftarrow min\{t \mid p+1 \le t \le p+d \; \& \; a(\gamma_t) \le p\}$;

  5       **if** $q \ne \infty$ **then**

  6         $\gamma_p \leftrightarrow \gamma_q$;

  7         $p \leftarrow q$;

  8       **else**

  9         **return** $\gamma'$;

10   **return** $\gamma$;

**Figure 4.1:** The Offline Optimal Packet Scheduling Algorithm.

### 4.2.2   Online Dominance Properties

The online algorithms exploit two dominance properties that may improve solution quality and performance. They select optimal solutions more adapted to online use and consider only a subset of the packets for scheduling at time $t$.

**Refining Optimal Solutions**   The offline problem typically has many optimal solutions that can be obtained by swapping a number of packets. These optimal solutions are not equivalent when used in an online setting. Figure 4.2 presents an enhanced optimization algorithm that favors those optimal solutions where high-weight packets are scheduled early. As discussed earlier, the offline algorithm schedules the packets from right to left and thus has a tendency to schedule packets of high-weight packets late. To overcome this side effect, the enhanced algorithm includes a postprocessing step (line 11) to favor high-weight packets early in the schedule and, in case of ties, packets with early deadlines. The postprocessing step, also depicted in figure 4.2, iterates over all pairs $(p, q)$ of positions ($p < q$) and swaps $\gamma_p$ with $\gamma_q$ if the swap is feasible and if

$$\langle w(\gamma_p), a(\gamma_p) \rangle < \langle w(\gamma_q), a(\gamma_q) \rangle,$$

that is, if packet $\gamma_p$ has a smaller weight than $\gamma_q$ or if they have the same weight but packet $\gamma_q$ has an earlier deadline.

   The enhanced algorithm thus schedules more valuable and less recent packets early without sacrificing quality or imposing additional constraints. As a consequence, it avoids the risk of losing high-quality packets and increases flexibility in the online decision process. The benefits of adapting optimization algorithms to an online setting is one of the themes of this book and we will come back to this issue in the context of vehicle dispatching.

**Dominated Packets**   The online algorithms for packet scheduling will restrict attention to a subset of the packets for the decision at time $t$. More precisely, the online algorithms consider at most one packet from each class and, for a class $c$, select the packet with the earliest deadline. This restriction does not decrease the quality of the schedule and it enables the algorithm to avoid losing earlier packets.

THEOREM 4.1   Let $A$ be the set of packets available at time $t$. For each class $c \in C$, an online algorithm needs to consider only a single packet $p \in A \cap c$, if one exists. Such a packet $p$ satisfies

$$\forall r \in A : c(p) = c(r) \rightarrow a(p) \leq a(r).$$

PROOF:   Consider a class $c$ and two packets $p, r \in C$ such that $a(r) \geq a(p)$. Let $\gamma^r$ be a solution where packet $r$ was scheduled at time $t$. We show that there exists a solution $\gamma^p$ with the same

ALGORITHM $\mathcal{O}(\langle R_1, \ldots, R_h \rangle)$

  1   $R \leftarrow \{r \in R_t \mid 1 \le t \le h\};$
  2   **for** $t \in H$ **do**
  3     $\gamma_t \leftarrow \bot;$
  4   **for** $r \in R$ ordered by decreasing $\langle w(r), a(r) \rangle$ **do**
  5     $p \leftarrow max\{t \in H \mid t \le a(r) + d \ \& \ \gamma_t = \bot\};$
  6     **if** $p \ne -\infty$ **then**
  7       **if** $p \ge a(r)$ **then**
  8         $\gamma_p \leftarrow r;$
  9       **else**
10        $\gamma \leftarrow \text{SHUFFLE}(\gamma, r, p);$
11   $\text{POSTPROCESS}(\gamma);$
12   **return** $\gamma;$

SHUFFLE$(\gamma, r, p)$

  1   $\gamma' \leftarrow \gamma;$
  2   $\gamma_p \leftarrow r;$
  3   **while** $p < a(r)$ **do**
  4     $q \leftarrow min\{t \mid p + 1 \le t \le p + d \ \& \ a(\gamma_t) \le p\};$
  5     **if** $q \ne \infty$ **then**
  6       $\gamma_p \leftrightarrow \gamma_q;$
  7       $p \leftarrow q;$
  8     **else**
  9       **return** $\gamma';$
10   **return** $\gamma;$

POSTPROCESS$(\gamma)$

  1   **for** $p \in H$ by increasing p **do**
  2    **for** $q \in H : p < q$ **do**
  3     **if** $\text{FEASIBLESWAP}(\gamma, p, q) \ \& \ \langle w(\gamma_p), a(\gamma_p) \rangle < \langle w(\gamma_q), a(\gamma_q) \rangle$ **then**
  4      $\gamma_p \leftrightarrow \gamma_q;$

FEASIBLESWAP$(\gamma, p, q)$

  1   **return** $a(\gamma_p) \le q \le a(\gamma_p) + d \ \wedge \ a(\gamma_q) \le p \le a(\gamma_q) + d.$

**Figure 4.2:** The Offline Optimal Packet Scheduling Algorithm Revisited.

weight where packet $p$ is scheduled at time $t$.

If $p \notin \gamma^r$, then $\gamma^p$ is obtained simply by replacing packet $r$ by packet $p$ since both are available at time $t$. Otherwise let $t'$ be the time when packet $p$ is scheduled. We have

$$a(p) \leq a(r) \leq t < t' \leq a(p) + d \leq a(r) + d.$$

As a consequence, $\gamma^p$ can be obtained by swapping packets $r$ and $p$. □

## 4.3 The Greedy Algorithm is Competitive

This section analyzes the greedy algorithm G, which consists of scheduling the packet with the highest weight at each step. The greedy algorithm is quite effective on packet scheduling and has a competitive ratio of two.

THEOREM 4.2 The greedy algorithm is 2-competitive for the packet scheduling problem.

PROOF: Let $\gamma^*$ be an optimal solution for a problem $\mathcal{P}$ and let $\gamma^g$ be the solution returned by algorithm G. We show that

$$\frac{w(\gamma^*)}{w(\gamma^g)} \leq 2.$$

Consider the set of packets $U$ scheduled in $\gamma^*$ but not in $\gamma^g$, that is,

$$U = \{ \gamma_t^* \mid t \in H \} \setminus \{ \gamma_t^g \mid t \in H \} \setminus \{\bot\}.$$

Consider a packet $p \in U$ scheduled at time $t$, that is, $\gamma_t^* = p$. Since $p \notin \gamma^g$, that is,

$$\nexists t \in H : \gamma_t^g = p$$

and since

$$a(p) \leq t \leq a(p) + d$$

by definition of a feasible schedule, packet $p$ was also available for selection at time $t$ for algorithm G. This implies that

$$w(\gamma_t^g) \geq w(p)$$

by definition of the greedy algorithm and hence

$$w(U) = \sum_{p \in U} w(p) \leq w(\gamma^g)$$

ALGORITHM REGRET($s_{t-1} : r, A, \gamma^*$)

1   $U \leftarrow A \setminus \{ \gamma_t^* \mid t \in H \}$;

2   **if** $r \notin \gamma^*$ **then**

3      **return** $\min(s \in [t, a(\gamma_t^*) + d]) \, w(\gamma_s^*) - w(r)$;

4   **else**

5      $t_r \leftarrow$ SELECT$(t \in H) \, \gamma_t^* = r$;

6      **if** READY$(\gamma_t^*, t_r)$ **then**

7         **return** 0;

8      **else**

9         **if** $\nexists p \in [t+1, a(\gamma_t^*) + d] : \; w(\gamma_p^*) \leq w(\gamma_t^*)$ **then**

10         **return** $w(\gamma_t^*) - \max(u \in U : \text{READY}(u, t_r)) \, w(u)$;

11         **else**

12            $t_s \leftarrow argmin(p \in [t+1, a(\gamma_t^*) + d] : w(\gamma_p^*) \leq w(\gamma_t^*)) \, w(\gamma_p^*)$;

13            **return** $w(\gamma_t^*)) - \max(u \in U \cup \{\gamma_{t_s}^*\} : \text{READY}(u, t_r)) \, w(u)$;

READY$(r, t)$

1   **return** $a(r) \leq t \leq a(r) + d$;

**Figure 4.3:** The Suboptimality Approximation for Packet Scheduling.

since at most one packet is scheduled at time $t$. As a consequence,

$$w(\gamma^*) \leq w(U) + w(\gamma^g) \leq 2w(\gamma^g)$$

and the result follows. $\qquad\qquad\qquad\qquad\qquad\qquad\qquad\qquad\qquad\qquad\qquad\qquad\square$

## 4.4  The Suboptimality Approximation

This section specifies the suboptimality approximation for packet scheduling that consists of swapping a constant number of packets in the optimal schedule and is based on a case analysis. Consider a packet $r \in$ FEASIBLE$(s_{t-1}, R_t)$ and let $A$ be the set of available packets (line 5 of CHOOSEREQUEST-R in figure 2.5). The algorithm computes the regret of packet $r$.

If packet $r$ is not scheduled (that is, $r \notin \gamma^*$), the algorithm tries rescheduling $\gamma_t^*$ instead of the packet of smallest weight in its time window (lines 2 and 3 in figure 4.3). The regret becomes

$$\min(s \in [t, a(\gamma_t^*) + d]) \, w(\gamma_s^*) - w(r),$$

since the replaced packet is removed from $\gamma^*$ and $r$ is added to the schedule. In the worst case, the replaced packet is $\gamma_t^*$ and the regret is $w(\gamma_t^*) - w(r)$.

If packet $r$ is scheduled at time $t_r$, the algorithm first tries to swap $r$ and $\gamma_t^*$ in which case the regret is 0 (lines 6 and 7 in figure 4.3). Note that packet $r$ is scheduled at time $t_r$ (line 5) and that $t_r$ is within the time window of packet $\gamma_t^*$ (line 6).

If the swap of packets $r$ and $\gamma_t^*$ is not possible, the algorithm tries to reschedule packet $\gamma_t^*$ in place of another packet $s$ of weight $w(s) \leq w(\gamma_t^*)$. If such a packet does not exist (line 9), the algorithm schedules at time $t_r$ the unscheduled packet of highest weight and the regret becomes

$$w(\gamma_t^*) - \max(u \in U : \text{READY}(u, t_r)) \ w(u)$$

where

$$\text{READY}(r, t) \equiv a(r) \leq t \leq a(r) + d$$

since packet $\gamma_t^*$ is lost. Otherwise, if packet $\gamma_t^*$ can be scheduled at time $t_s$ (lines 12 and 13), the algorithm concludes by selecting the best possible unscheduled packet that may be scheduled at time $t_r$ (line 13) and the regret becomes

$$w(\gamma_t^*) - \max(u \in U \cup \{\gamma_{t_s}^*\} : \text{READY}(u, t_r)) \ w(u).$$

This regret function takes $O(\max(d, |C|))$ time, which is sublinear in $|H|$ and $|R|$ and essentially negligible for this application. We now prove that it provides a 2-approximation.

> **THEOREM 4.3**  The regret algorithm for packet scheduling is a 2-approximation.

**PROOF:**  Let $r \in \text{FEASIBLE}(s_{t-1}, R_t)$ and let $A$ be the available packets (line 5 of CHOOSEREQUEST-R in figure 2.5). Let $\gamma^*$ be an optimal solution, that is,

$$\gamma^* = \mathcal{O}(\mathbf{S}_{t_1}, A),$$

let $\gamma^r$ be an optimal solution when $r$ is scheduled at time $t$, that is,

$$\gamma^r = \mathcal{O}(\mathbf{S}_{t-1} : r, A),$$

and let $\tilde{\gamma}^r$ be the solution obtained by the regret function. We show that

$$\frac{w(\gamma^r)}{w(\tilde{\gamma}^r)} \leq 2.$$

Most of the proof consists of showing that for each lost packet $l$ there is another packet in $\gamma^*$ whose weight is at least $w(l)$, giving us a 2-approximation since $w(\gamma^r) \leq w(\gamma^*)$. In the proof, we denote $\gamma_t$ by $x$ for simplicity.

First observe that the result holds when $w(x) \leq w(r)$ since, in the worst case, the regret function loses only packet $x$. So we restrict attention to $w(x) \geq w(r)$.

If $x \in \tilde{\gamma}^r$, that is, if the regret function swaps $x$ with another packet $y$ (lines 12 and 13 in figure 4.3), the result also holds since $w(y) \leq w(x)$ because of the condition in line 9.

If $x \notin \tilde{\gamma}^r$ and $x$ can be scheduled after time $t$, it means that there exists a packet $y$ at each of these times satisfying $w(y) \geq w(x)$ (line 9 again) and the result holds.

It thus remains to consider the case where $x$ can be scheduled only at time $t$ and is thus lost in $\gamma^r$. If $r \notin \gamma^*$, the regret function is optimal, since otherwise $r$ would be in the optimal schedule after time $t$. Otherwise, it is necessary to reason about a collection of packets. We have

$$w(\gamma^*) = w(x) + w(r) + w(S)$$

where

$$S = \{p \in \gamma^* \mid p \neq x \ \& \ p \neq y\}.$$

We also know that

$$w(\tilde{\gamma}^r) \geq w(r) + w(S)$$

since, in the worst case, the regret function loses packet $x$. Finally,

$$w(\gamma^r) = w(r) + w(Z)$$

where $Z$ are the packets scheduled after time $t$. Since $\gamma^*$ is optimal, it follows that

$$w(Z) \leq w(r) + w(S)$$

and hence

$$w(\gamma^r) = w(r) + w(Z) \leq 2w(r) + w(S) \leq 2w(\tilde{\gamma}^r). \qquad \square$$

## 4.5   Experimental Setting

This section describes the setting for the experimental results. In the online problem, the packets are revealed dynamically at each time step according to the distributions originally specified in [28] and also used in [13]. In particular, arrival distributions are specified by independent Markov Models (MMs), one for each packet class. Each MM consists of three states corresponding to sparse, medium, and dense arrival patterns, as shown in figure 4.4. There exist transitions from sparse to medium, medium to dense, and dense to sparse, as well as self-transitions. The self-transition for each state is drawn uniformly from the interval [0.9, 1.0]. The probability that a packet arrives at a given time step is drawn uniformly from [0.0, 0.1] when the MM is in the sparse state, from [0.2,

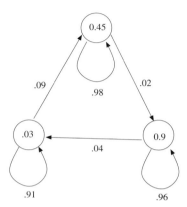

**Figure 4.4:** An Example of Markov Model for Packet Scheduling.

0.5] in the medium state, and from [0.7, 1.0] in the dense state. The so-obtained MMs are then normalized so that the expected number of packets per time step is 1.5. The deadline $d$ for the packets in these instances is 20.

The evaluation considers problems with as many as thirty packet classes. These instances were generated to balance low-, medium-, and high-profit classes, and table 4.1 describes the weights of the packet classes. The "rank" column indicates which packet classes are used for a problem with $\rho$ packets classes, that is, a $\rho$-class problem uses packets with rank no greater than $\rho$. All results described in this chapter are for problems consisting of 200,000 times steps.

## 4.6   Experimental Results

This section presents the experimental results for packet scheduling. It compares the online algorithms and studies their scalability and their robustness. It also reviews the impact of the sampling, since it is unrealistic to sample the future for the entire horizon in this application. The effect of precomputations on the quality of the decisions is also examined. Finally, this section studies the anticipativity assumption in packet scheduling.

### 4.6.1   Comparison of the Algorithms

Figure 4.5 depicts the average packet loss as a function of the number of available optimizations $\mathcal{T}$ for the various algorithms on the 7-class and 14-class problems. It also gives the optimal, a posteriori, packet loss ($O$), that is, the offline solution for the instances after the uncertainty has been completely revealed. The results are the average of 10 runs on 10 different instances with a sampling horizon $f$ of 50. In other words, the online algorithms sample the future using a call

| Low | | Medium | | High | |
|---|---|---|---|---|---|
| Weight | Rank | Weight | Rank | Weight | Rank |
| 5 | 1 | 600 | 2 | 1000 | 3 |
| 10 | 4 | 800 | 5 | 2000 | 6 |
| 20 | 7 | 500 | 8 | 1500 | 9 |
| 1 | 10 | 700 | 11 | 1200 | 12 |
| 15 | 13 | 750 | 14 | 3000 | 15 |
| 50 | 16 | 550 | 17 | 2500 | 18 |
| 30 | 19 | 650 | 20 | 3500 | 21 |
| 25 | 22 | 400 | 23 | 4000 | 24 |
| 40 | 25 | 450 | 26 | 5000 | 27 |
| 3 | 28 | 575 | 29 | 4500 | 30 |

**Table 4.1:** The Packet Classes Used in the Experimental Results.

CALL$(t, f)$, thus returning sequences of 50 sets of future packets. Recall that the greedy algorithm G is a 2-competitive algorithm so that the online stochastic algorithms are compared with a good online algorithm.

The experimental results on the 7-class problems indicate the value of stochastic information as algorithm $\mathcal{E}$ outperforms the oblivious algorithms G and LO and bridge much of the gap between these algorithms and the optimal solution. LO is worse than G, illustrating the (frequent) pathological behavior of over-optimization in online settings. The experimental results also indicate that algorithm $\mathcal{C}$ outperforms $\mathcal{E}$ whenever few optimizations are available. The improvement is particularly significant when there are very few available optimizations. Consensus is dominated by algorithm $\mathcal{E}$ when the number of available optimizations increases, although it still produces significant improvements over the oblivious algorithms. This is pertinent, since algorithm $\mathcal{E}$ is not practical for many problems with time constraints, including the dispatching and routing problems discussed in chapters 9 and 10.

The benefits of the regret algorithm $\mathcal{R}$ are apparent. Algorithm $\mathcal{R}$ dominates all the other algorithms including algorithm $\mathcal{C}$ when there are very few offline optimizations (strong time constraints) and algorithm $\mathcal{E}$ even when there are a reasonably large number of them (weak time constraints). Observe that algorithm $\mathcal{R}$ with 10 optimizations produces the same solution quality as algorithm $\mathcal{E}$ with 50 optimizations. Since the number of feasible requests at each time $t$ is about 5 in average, the experimental results agree nicely with the theoretical analysis.

On 14-class problems, the algorithms exhibit similar behaviors. Moreover, it takes more optimizations for algorithm $\mathcal{E}$ to dominate algorithm $\mathcal{C}$ and to achieve the same quality as algorithm $\mathcal{R}$. Indeed algorithm $\mathcal{E}$ now needs about 20 optimizations per step to improve the quality of consensus

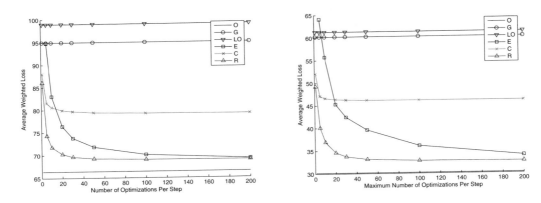

**Figure 4.5:** Comparisons of the Algorithms on Packet Scheduling: 7-Class and 14-Class Problems.

and about 200 optimizations to reach the same quality as algorithm $\mathcal{R}$. Section 4.6.2 studies the impact of the number of packet classes in more detail.

Figure 4.6 depicts the results of the 10 instances of 7-class problems. The results show that the behavior of the algorithms differs little, the main differences being exactly when algorithm $\mathcal{E}$ outperforms $\mathcal{C}$ and when it becomes similar in quality to algorithm $\mathcal{R}$. In general, algorithm $\mathcal{E}$ needs between 100 and 200 optimizations to reach the same quality as $\mathcal{R}$ and about 10 to 20 optimizations to dominate $\mathcal{C}$.

## 4.6.2 Scalability

This section studies the impact of the number of packet classes on the behavior of the algorithms. It compares the algorithms for instances with 7 to 30 classes when $\mathcal{T} = 30$ and $\Delta = 50$. Figure 4.7 depicts the average loss at each time step as the number of classes is increased, as well as the same results as a percentage of the online solution over the optimal offline solution.

As expected, the performance of $\mathcal{E}$ is most affected by the increase in number of classes. In the 7-class problems, $\mathcal{E}$ is roughly 10 percent worse than optimal, which increases to as much as 40 percent as the number of classes increases. This is in sharp contrast to $\mathcal{C}$, which merely increases from 20 percent to 40 percent and matches the performance of $\mathcal{E}$ for problems with many classes. Perhaps the most interesting result is provided by algorithm $\mathcal{R}$: its performance degradation with respect to the optimal offline solution is negligible as the number of classes increases. These results demonstrate that, on packet scheduling, algorithm $\mathcal{R}$ (and to a lesser extent algorithm $\mathcal{C}$) is more scalable than algorithm $\mathcal{E}$ when the number of packet classes increases.

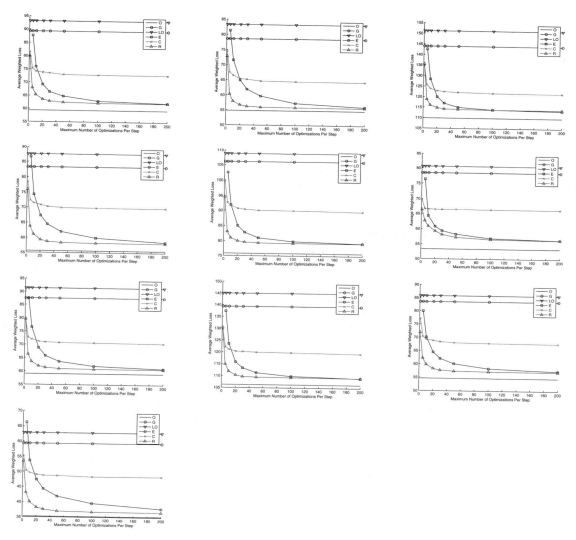

**Figure 4.6:** Comparisons of the Algorithms on Packet Scheduling: 7-Class Instances.

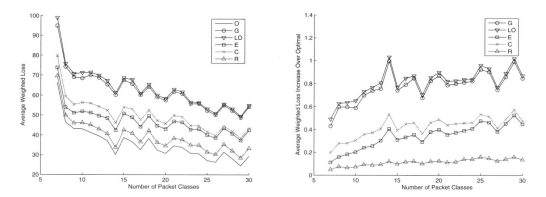

**Figure 4.7:** The Impact of the Number of Packet Classes in Packet Scheduling.

### 4.6.3   The Impact of Noisy Distributions

It is legitimate to ask how the algorithms behave when the exact distribution is not known. This section looks at how the online stochastic algorithms behave under noisy distributions (chapter 11 looks at how to learn distributions online). Figure 4.8 depicts the behavior of the algorithms when there are fewer or more packets than predicted by the distribution when $\mathcal{T} = 50$. More precisely, the figure considers various noisy distributions ranging from the case where there are 25 percent fewer packets to the case where there are 25 percent more packets. The results indicate that the algorithms remain effective under these conditions. They also seem to indicate that it is better to assume too many packets than too few, since the quality degradation is smaller in this case. Figures 4.9 and 4.10 elaborate on these results and depict the behavior of the algorithms under various noisy distributions. The results show that the algorithms continue to produce high-quality results under these conditions. The results also demonstrate that, under this noise model, the behavior of the algorithms is less affected when the distribution predicts more packet arrivals.

### 4.6.4   The Impact of the Sampling Horizon

In this application consisting of 200,000 steps, it is unrealistic to sample the entire time horizon to make a decision. The experimental results presented so far used a sampling horizon of 50. This section evaluates the impact of this parameter.

Figure 4.11 depicts the impact of the sampling horizon $\Delta$ for the 7-class instances when $\mathcal{T}$ is fixed at 20. The results indicate that the choice $\Delta = 50$ is excellent for $\mathcal{E}$. Increasing $\Delta$ beyond 50 provides limited benefits for algorithms $\mathcal{E}$ and $\mathcal{R}$, while shorter time horizons decrease their effectiveness. The effectiveness of algorithm $\mathcal{C}$ can be boosted by increasing the sampling horizon.

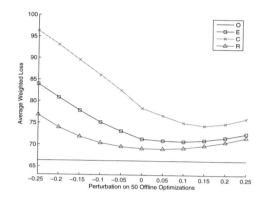

**Figure 4.8:** The Impact of Noisy Distributions in Packet Scheduling.

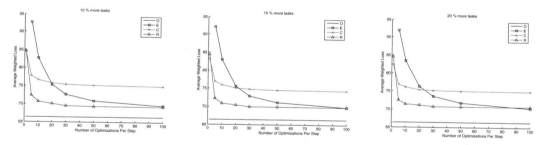

**Figure 4.9:** The Impact of Noisy Distributions in Packet Scheduling: Predicting Too Many Packets.

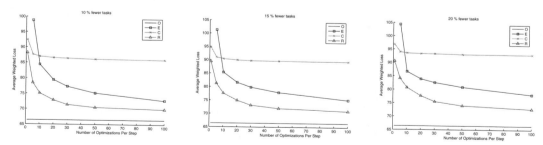

**Figure 4.10:** The Impact of Noisy Distributions in Packet Scheduling: Predicting Too Few Packets.

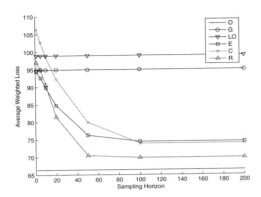

**Figure 4.11:** The Impact of the Sampling Horizon on Packet Scheduling for $\mathcal{T} = 20$.

Indeed when $\Delta$ is fixed at 100, the point at which $\mathcal{E}$ dominates $\mathcal{C}$ shifts from 20 to 40 optimizations.

### 4.6.5 The Impact of Precomputation

It is important to examine the impact of precomputation on the quality of the algorithms for packet scheduling. Precomputation is critical in applications in which decisions must be taken immediately, but it may introduce a bias in the sampling if scenarios are preserved across decisions, which is often necessary in the presence of tight time constraints. Indeed the old scenarios may introduce noise with respect to the exact distribution, since transitions may have taken place in the Markov models.

Figure 4.12 depicts experimental results on precomputation. It reports the solution quality of the algorithms when scenario solutions are preserved until they become inconsistent with the decisions. The results indicate the benefits of keeping old solutions under extreme conditions, that is, when there is time for fewer than five optimizations.[1] However, the results also show that the inconvenience of preserving old, noisy scenarios outweighs their benefits under milder time constraints. As a result, precomputation should be used carefully, since the benefits of keeping old scenarios may be outweighed by the noise they introduce. Of course, in practice, there may be no alternative to using precomputation.

## 4.7 The Anticipativity Assumption

We now study the anticipativity assumption for the packet scheduling instances considered in this chapter. Recall that, informally speaking, the anticipativity assumption implies that the same

---

[1]Such extreme conditions arise in the vehicle routing applications in chapter 10.

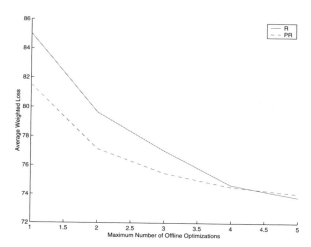

**Figure 4.12:** The Impact of Precomputation on Packet Scheduling.

packet is scheduled first in all scenarios at time $t$.

Figure 4.13 provides some experimental data on the "validity" of the anticipativity assumption. It depicts how many times (in percentage) the packet selected by algorithm $\mathcal{R}$ is scheduled first in the various scenarios. The figure reports the average, minimum, and maximum number of times (in percentage) the selected packet is scheduled first as a function of the number of optimizations per step. It also reports the average plus or minus the standard deviation. The results indicate that, on average, the selected packet appears first in about 90 percent of the scenarios (depending on the number of optimizations). The standard deviation is always less than 20 percent. In other words, the anticipativity assumption largely holds in these packet scheduling problems (which were originally proposed in [28]).

Figure 4.14 gives some insight into why the anticipativity assumption is largely valid in the packet scheduling instances. The figure displays the packet selection of algorithm $\mathcal{R}$ as a function of its relative deadline. In the figure, a relative deadline of zero means that the packet is about to expire, while a relative deadline of twenty means that the packet just arrived. The figure also distinguishes between greedy selections (the selected packet is also the greedy choice) and nongreedy selections. The figure reveals several properties of the algorithm behavior.

1. The most recent packets are selected the most often, but this selection is almost always greedy.

2. The packets with relative deadlines 0 to 9 are selected about the same number of times. The order of these packets is not really significant. They differ in the kind of selection they correspond to. In particular, the packets about to expire are almost never greedy selections.

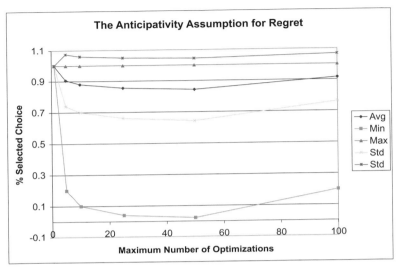

**Figure 4.13:** The Anticipativity Assumption for Algorithm $\mathcal{R}$ in Packet Scheduling.

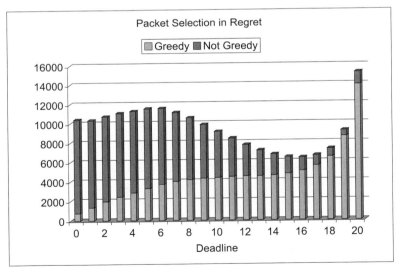

**Figure 4.14:** Relative Deadline Selection for Algorithm $\mathcal{R}$ in Packet Scheduling.

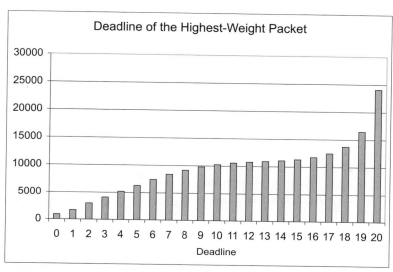

**Figure 4.15:** Relative Deadline of the Highest-Weight Packet in Algorithm $\mathcal{R}$.

This is in sharp contrast to the most recent packets (deadlines 18 to 20).

3. The packets with relative deadlines 10 to 19 are selected less often than those with deadlines 0 to 9. But, once again, there is not much difference between many of these packets, indicating that their order is not that significant either.

4. The greedy nature of the selection monotonically increases from 0 to 20, with a steep increase in the most recent packets and, in particular, for the packets just arrived.

Figure 4.15 gives some additional insight into the behavior of the algorithm: it plots the deadline of the highest-weight packet (that is, what the greedy choice would be) as a function of its relative deadline during the execution of algorithm $\mathcal{R}$. The figure shows that the packets of highest weight are concentrated mostly in recent arrivals. Packets about to expire rarely have the highest weight and are selected by algorithm $\mathcal{R}$ because they are expected to be more valuable than future packets.

In summary, the anticipativity assumption is largely valid for algorithm $\mathcal{R}$ in the packet scheduling instances. Intuitively, the main reason is that either the choice is greedy, which means that one of the most recent packets is selected in almost all scenarios, or the choice is not, in which case the order of the packets does not seem significant since the number of selections for the various deadlines do not differ significantly.

Figures 4.16, 4.17, and 4.18 depict the same information for algorithm $\mathcal{C}$. The results indicate that, in the average, the selected packet appears first in about 70 percent of the scenarios (depending on

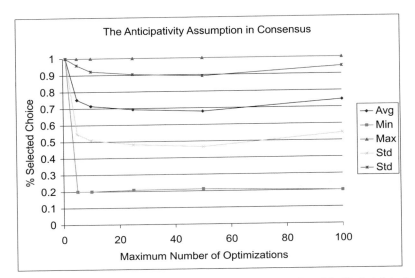

**Figure 4.16:** The Anticipativity Assumption for Algorithm $\mathcal{C}$ in Packet Scheduling.

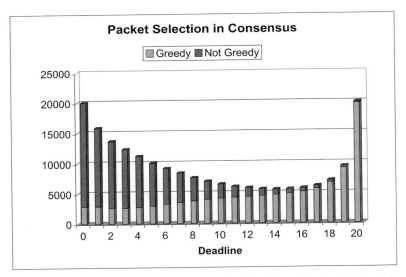

**Figure 4.17:** Relative Deadline Selection for Algorithm $\mathcal{C}$ in Packet Scheduling.

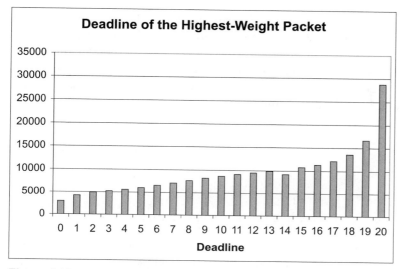

**Figure 4.18:** Relative Deadline of the Highest-Weight Packet in Algorithm $\mathcal{C}$.

the number of optimizations). The anticipativity assumption is still largely valid, although there is a significant gap compared to algorithm $\mathcal{R}$. Figure 4.17 illustrates some of the differences between the behaviors of algorithms $\mathcal{R}$ and $\mathcal{C}$ and why there is such a gap between them. In particular, algorithm $\mathcal{C}$ now exhibits two peaks, one for the greedy choice with relative deadline 20 and one for the nongreedy choice of relative deadline 0. In other words, algorithm $\mathcal{R}$ seems to anticipate the future better and avoids having to so often select a packet about to expire. There are also greater variations across the deadlines in the number of selections. Figure 4.17, which plots the deadline of the highest-weight packet, resembles its regret counterpart but is not as smooth.

It should be emphasized, however, that not all distributions for packet scheduling are anticipative.

**Example 4.1** Consider the horizon $[1, 2d]$ and the following two scenarios [60]. In both scenarios, $d$ packets of weight 1 arrive at time 1, and 1 packet of weight $\omega > 1$ arrives at each time step $t$ until time $d$ (that is, $1 \leq t \leq d$). In addition, in the first scenario, $d$ packets of weight $\omega$ arrive at step $d + 1$, while no packets arrive in the second scenario. The optimal solution for the first scenario is

$$\omega, \ldots, \omega, \omega, \ldots, \omega$$

with a weight of $2d\omega$, while the optimal solution for the second scenario is

$$1, \ldots, 1, \omega, \ldots, \omega$$

with a total weight of $d + d\omega$. The distribution is not anticipative for packet scheduling since it schedules different packets at time 1. It is not $k$-anticipative for any $k$, in fact, since it is only when the uncertainty is revealed at time $d+1$ that we will know what the optimal solution for the specific scenario at time 1 is. Algorithms $\mathcal{R}$ and $\mathcal{E}$ systematically choose packets of weight $\omega$ giving an optimal solution of weight $2d\omega$ in the first scenario and a suboptimal solution of weight $d\omega$ in the second scenario. The expected weight of the algorithms is thus

$$\frac{3d\omega}{2}$$

while the expected optimal weight is

$$\frac{3d\omega + d}{2}.$$

It is possible to scale up these scenarios by repeating these $2d$ steps multiple times.

In the above example, the algorithms have no way to make the right decisions at times 1 to $d$ because they cannot differentiate the scenarios until time $d + 1$ and the optimal solutions to these scenarios make distinct choices. This is not a limitation of algorithms $\mathcal{E}$ and $\mathcal{R}$: any online algorithm will suffer the same fate. Rather this is a property of the distribution that prevents the algorithms from differentiating the scenarios early enough. As a consequence, it is fair to conclude that the anticipativity assumption captures a fundamental property of well-behaved distributions.

Figure 4.19 provides some other insight into the anticipativity assumption. It compares at each time step the optimal value of the scenarios and compares it to the optimal offline for the same time windows. The top figure describes the average optimal value, while the bottom figure depicts the standard deviation over time. The results show that the optimal values of the scenarios are not too optimistic and that increasing the number of scenarios significantly reduces the variance of the optimal value.

## 4.8   Notes and Further Reading

Reference [60] is a beautiful paper on packet scheduling. It discusses the motivation and relevance of packet scheduling, presents several models, and studies a variety of online algorithms. In particular, it analyzes the greedy algorithm and a form of local optimization. The proof that the greedy algorithm is 2-competitive is given in that paper. Example 4.1 is also taken from that paper where it is used to prove a lower bound on the quality of any online algorithm. Algorithm $\mathcal{E}$ was used in [28] for packet scheduling. The reading of that paper led us to reframe for packet scheduling algorithm $\mathcal{C}$ initially developed for online stochastic vehicle routing [12]. It led to the development of the generic online scheduling problems presented in [10] and to the regret algorithm $\mathcal{R}$ [11].

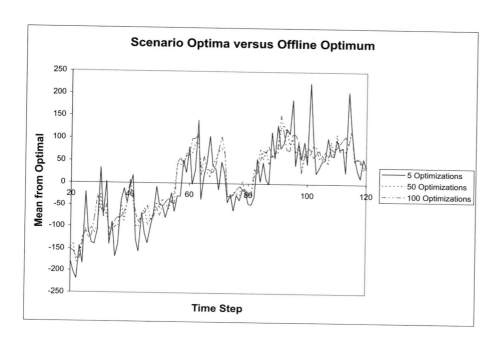

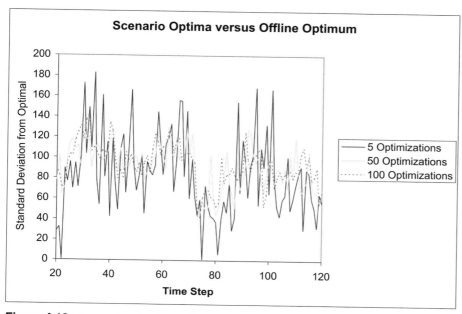

**Figure 4.19:** Comparing the Optimal Solutions of $\mathcal{R}$ with the Offline Optimum over Time.

# II ONLINE STOCHASTIC RESERVATIONS

# 5 Online Stochastic Reservations

*My mother's menu consisted of two choices: Take it or leave it.*
— Buddy Hackett

This chapter considers online reservation problems that share many aspects with online scheduling problems. However, they differ in that the goal is not to select which request to serve at time $t$. Rather the decision is about how to serve a request and whether the request must be served at all. This chapter specifies offline and online reservation systems and adapts algorithms $\mathcal{E}$, $\mathcal{C}$, and $\mathcal{R}$ to this new setting. It also shows how to accommodate request cancellations. Sections 5.1 and 5.2 specify the offline and online reservation problems. Section 5.3 presents the generic online algorithm, which is then instantiated to the expectation, consensus, and regret algorithms. Section 5.7 shows how to accommodate cancellations.

## 5.1 The Offline Reservation Problem

The offline reservation problem is defined in terms of $n$ bins $B$ and each bin $b \in B$ has a capacity $C_b$. It receives as input a set $R = \{r_1, \ldots, r_h\}$ of requests. Each request is characterized typically by its capacity and its reward, which may or may not depend on which bin the request is allocated to. The goal is to find an assignment of a subset $T \subseteq R$ of requests to the bins satisfying the problem-specific constraints $C$ and maximizing the problem-specific objective function $w$.

**Formalization** To formalize the online algorithms concisely, it is convenient to assume the existence of a dummy bin $\perp$ with infinite capacity. The rejected requests all can be assigned to the dummy bin, allowing rejection and allocation decisions to be modeled similarly. A solution $\sigma$ can be seen as an assignment from $R$ to $B_\perp$, where $B_\perp$ denotes $B \cup \{\perp\}$. The objective function is specified as a function $w$ over assignments and the problem-specific constraints as a relation over assignments. The offline reservation problem can be formalized as

$$\max_\sigma \quad w(\sigma)$$
$$\text{such that}$$
$$C(\sigma),$$

where $\sigma$ range over all assignments from $R$ to $B_\perp$.

**Notations** This chapter and the next heavily manipulate assignments and use the following notations. If $\sigma$ is an assignment from $R$ to $B_\perp$ and $r \in R$, $\sigma(r)$ denotes the assignment of $r$. If $r \in R$ and

$b \in B_\perp$, $\sigma[r \leftarrow b]$ denotes the assignment $\sigma$ where $r$ is assigned to bin $b$ and all other assignments remain the same, that is,

$$\begin{aligned}
\sigma[r \leftarrow b](r) &= b \\
\sigma[r \leftarrow b](r') &= \sigma(r') \quad \text{if } r' \neq r.
\end{aligned}$$

The notation

$$\sigma[r_1 \leftarrow b_1; \ldots; r_k \leftarrow b_k]$$

is an abbreviation for

$$\sigma[r_1 \leftarrow b_1] \ldots [r_k \leftarrow b_k].$$

The assignment $\sigma_\perp$ is the empty assignment that satisfies

$$\forall r \in R : \sigma(r) = \perp.$$

The assignment $\sigma \downarrow R$ denotes the assignment $\sigma$ where the requests in $R$ are now assigned to the dummy bin, that is,

$$\begin{aligned}
(\sigma \downarrow R)(r) &= \perp \quad \text{if } r \in R \\
(\sigma \downarrow R)(r) &= \sigma(r) \quad \text{if } r \notin R.
\end{aligned}$$

Finally, we also use $\mathcal{F}(r, \sigma)$ to denote the set of feasible allocations for request $r$ given the assignment $\sigma$ that is,

$$\mathcal{F}(r, \sigma) = \{ \, b \mid C(\sigma[r \leftarrow b]) \, \}.$$

## 5.2   The Online Problem

The online reservation problem assumes, once again, a discrete model of time and considers a time horizon $H = [1, h]$. In the online problem, the set of requests $R$ is not known a priori but is revealed online, one request at a time, as the algorithm executes. At each time $t \in H$, the request $r_t$ is revealed and the online algorithm has at its disposal the assignment $\sigma_{t-1}$ that contains the allocations for requests $r_1, \ldots, r_{t-1}$. The goal of the online algorithm is to decide the allocation of request $r_t$, thus producing an assignment $\sigma_t$ satisfying the problem-specific constraint $C(\sigma_t)$. In the online problem, the sequence $\langle r_1, \ldots, r_h \rangle$ of requests is drawn from a distribution $\mathcal{I}$. In other words, $\langle r_1, \ldots, r_h \rangle$ can be seen as the realizations of random variables $\langle \xi_1, \ldots, \xi_h \rangle$ whose distribution is specified by $\mathcal{I}$. Once again, the online algorithms have at their disposal a procedure to solve, or approximate, the offline problem, and a procedure to sample the distribution $\mathcal{I}$.

It is interesting to contrast the online reservation problem with the online scheduling problem. In online scheduling, the key issue is which request to serve at each step; requests that are not selected remain available for future selections. In online reservation, the key issue is whether to serve the incoming request, and how. Moreover, whenever a request is accepted, it must be assigned a specific bin and the algorithm is not allowed to reshuffle the assignments later.

ONLINE ALGORITHM $\mathcal{A}(\langle r_1, \ldots, r_h \rangle)$
1   $\sigma_0 \leftarrow \sigma_\perp$;
2   **for** $t \in H$ **do**
3       $b \leftarrow$ CHOOSEALLOCATION$(\sigma_{t-1}, r_t)$;
4       $\sigma_t \leftarrow \sigma_{t-1}[r_t \leftarrow b]$;
5   **return** $\sigma_h$;

**Figure 5.1:** The Generic Online Reservation Algorithm.

## 5.3   The Generic Online Algorithm

The algorithms for online reservations are all instances of the generic schema depicted in figure 5.1; they differ only in how they implement function CHOOSEALLOCATION in line 3. The online algorithm starts with an empty allocation (line 1). At each decision time $t$, the online algorithm considers the current allocation $\sigma_{t-1}$ and the current request $r_t$. It chooses the bin $b$ to allocate the request (line 3), which is then included in the new assignment $\sigma_t$ (line 4). The algorithm returns the last assignment $\sigma_h$ whose value is $w(\sigma_h)$ (line 5). The goal of the online algorithm is to maximize the expected profit

$$\mathbb{E}_{\xi_1, \ldots, \xi_h} [\mathcal{A}(\langle \xi_1, \ldots, \xi_h \rangle)].$$

For this purpose, it has two main black-boxes at its disposal:

1. an optimization algorithm $\mathcal{O}$ that, given an allocation $\sigma_{t-1}$ of requests $r_1, \ldots, r_{t-1}$ and a set of requests $R$, returns an optimal solution

$$\mathcal{O}(\sigma_{t-1}, R)$$

   to the offline problem consistent with the past decisions, that is,

$$\max_{\sigma} \quad w(\sigma)$$
$$\text{such that}$$
$$C(\sigma);$$
$$\forall i \in 1..t-1 : \sigma(r_i) = \sigma_{t-1}(r_i).$$

2. a conditional sampling procedure SAMPLE that, given a time $t$ and a sampling horizon $f$, generates a sequence

$$\text{SAMPLE}(t, f) = \langle r_{t+1}, \ldots, r_f \rangle$$

   of realizations for the random variables $\xi_{t+1}, \ldots, \xi_f$ obtained by sampling the distribution $\mathcal{I}$.

CHOOSEALLOCATION-$\mathcal{E}(\sigma_{t-1}, r_t)$
1   $F \leftarrow \mathcal{F}(\sigma_{t-1}, r_t)$;
2   **for** $b \in F$ **do**
3      $f(b) \leftarrow 0$;
4   **for** $i \leftarrow 1 \ldots \mathcal{T}/|F|$ **do**
5      $A \leftarrow$ SAMPLE$(t, h)$;
6      **for** $b \in F$ **do**
7         $f(b) \leftarrow f(b) + w(\mathcal{O}(\sigma_{t-1}[r_t \leftarrow b], A))$;
8   **return** $argmax(b \in F) \; f(b)$;

**Figure 5.2:** The Expectation Algorithm for Online Stochastic Reservation.

The online stochastic reservation problem shares the same properties as online stochastic scheduling. There is no recourse and, as soon as a request is realized, its data is fully revealed. The decision at time $t$ is constrained by past allocations and has an impact on future allocations. It also can be viewed as an approach to the multistage stochastic problem

$$\mathbb{E}_{\xi_1} \max_{b_1 \in \mathcal{F}(r_1, \sigma_0)} \cdots \mathbb{E}_{\xi_h} \max_{b_h \in F(r_h, \sigma_{h-1})} w(\sigma_0[\xi_1 \leftarrow b_1, \ldots, \xi_h \leftarrow b_h]).$$

## 5.4  The Expectation Algorithm

The online anticipatory algorithm $\mathcal{E}$ is presented in figure 5.2. Line 1 computes the set $F$ of feasible allocations for request $r_t$. Lines 2 and 3 initialize the evaluation $f(b)$ of every feasible allocation $b$. The algorithm then generates $\mathcal{T}/|F|$ scenarios (lines 4 and 5). For each scenario, it successively considers each feasible allocation $b$ (line 6). For each such allocation $b$, it schedules $r_t$ in bin $b$ and applies the optimization algorithm using the sampled requests $A$ (line 7). The evaluation of bin $b$ is incremented in line 7 with the weight of the optimal assignment $\sigma^*$. Once all the bin allocations are evaluated over all scenarios, the algorithm returns the bin $b$ with the highest evaluation. Note that the algorithm does not keep track of past requests: the accepted requests are present in the allocation, while the rejected requests are no longer relevant.

## 5.5  The Consensus Algorithm

Figure 5.3 depicts the consensus algorithm. Algorithm $\mathcal{C}$ solves every sample once, using $\mathcal{T}$ samples instead of $\mathcal{T}/|F|$. Observe the optimization call in line 6 that uses $\sigma_{t-1}$ and $A \cup \{r_t\}$. It produces an optimal allocation for $r_t$ for the scenario $A$ and this allocation is credited one unit in line 7.

CHOOSEALLOCATION-$\mathcal{C}(\sigma_{t-1}, r_t)$
1   $F \leftarrow \mathcal{F}(\sigma_{t-1}, r_t)$;
2   **for** $b \in F$ **do**
3      $f(b) \leftarrow 0$;
4   **for** $i \leftarrow 1 \ldots \mathcal{T}/|F|$ **do**
5      $A \leftarrow$ SAMPLE$(t, f)$;
6      $\sigma^* \leftarrow \mathcal{O}(\sigma_{t-1}, A \cup \{r_t\})$;
7      $f(\sigma^*(r_t)) \leftarrow f(\sigma^*(r_t)) + 1$;
8   **return** $argmax(b \in F)\ f(b)$;

**Figure 5.3:** The Consensus Algorithm for Online Stochastic Reservation.

CHOOSEALLOCATION-$\mathcal{R}(\sigma_{t-1}, r_t)$
1   $F \leftarrow \mathcal{F}(\sigma_{t-1}, r_t)$;
2   **for** $b \in F$ **do**
3      $f(b) \leftarrow 0$;
4   **for** $i \leftarrow 1 \ldots \mathcal{T}/|F|$ **do**
5      $A \leftarrow$ SAMPLE$(t, f)$;
6      $\sigma^* \leftarrow \mathcal{O}(\sigma_{t-1}, A \cup \{r_t\})$;
7      $f(\sigma^*(r_t)) \leftarrow f(\sigma^*(r_t)) + w(\sigma^*)$;
8      **for** $b \in F \setminus \{\sigma^*(r_t)\}$ **do**
9         $f(b) \leftarrow f(b) + w(\widetilde{\mathcal{O}}(\sigma_{t-1}[r_t \leftarrow b], A, \sigma^*))$;
10  **return** $argmax(b \in F)\ f(b)$;

**Figure 5.4:** The Regret Algorithm for Online Stochastic Reservation.

## 5.6  The Regret Algorithm

The regret algorithm $\mathcal{R}$ is the recognition that, in many applications, it is possible to estimate the loss of suboptimal allocations quickly. In other words, once the optimal solution $\sigma^*$ of a scenario is computed, algorithm $\mathcal{E}$ can be approximated with one optimization. The definitions can be easily generalized from online scheduling to online reservation. For instance, the next definition captures the concept of suboptimality approximation for online reservation.

**Definition 5.1 (Suboptimality Approximation)** Let $\mathcal{P}$ be a reservation problem and $\mathcal{O}$ be an offline algorithm for $\mathcal{P}$. Algorithm $\widetilde{\mathcal{O}}$ is a suboptimality approximation for $\mathcal{P}$ if, for every pair

ONLINE ALGORITHM $\mathcal{A}(r_h, K_h)$

1  $\sigma_0 \leftarrow \sigma_\perp$;
2  **for** $t \in H$ **do**
3  $\quad \sigma_{t-1} \leftarrow \sigma_{t-1} \downarrow K_t$;
4  $\quad b \leftarrow$ CHOOSEALLOCATION$(\sigma_{t-1}, r_t)$;
5  $\quad \sigma_t \leftarrow \sigma_{t-1}[r_t \leftarrow b]$;
6  **return** $\sigma_h$;

**Figure 5.5:** The Generic Online Reservation Algorithm with Cancellations.

CHOOSEALLOCATION-$\mathcal{C}(\sigma_{t-1}, r_t)$

1  $F \leftarrow \{ b \in B_\perp \mid C(\sigma_{t-1}[r_t \leftarrow b]) \}$;
2  **for** $b \in F$ **do**
3  $\quad f(b) \leftarrow 0$;
4  **for** $i \leftarrow 1 \ldots \mathcal{T}/|F|$ **do**
5  $\quad (A, K) \leftarrow$ SAMPLE$(t, f)$;
6  $\quad \sigma^* \leftarrow \mathcal{O}(\sigma_{t-1} \downarrow K, A \cup \{r_t\} \setminus K)$;
7  $\quad f(\sigma^*(r_t)) \leftarrow f(\sigma^*(r_t)) + 1$;
8  **return** $argmax(b \in F)\ f(b)$;

**Figure 5.6:** The Consensus Algorithm for Online Stochastic Reservation with Cancellation.

$(\sigma_{t-1}, r_{t..h})$ and every bin $b \in$ FEASIBLE$(\sigma_{t-1}, r_{t..h})$,

$$w(\mathcal{O}(\sigma_{t-1}[r_t \leftarrow b], r_{t..h})) \leq \beta\ w(\widetilde{\mathcal{O}}(\sigma_{t-1} : r, r_{t..h}, \mathcal{O}(\sigma_{t-1}, r_{t..h})))$$

for some constant $\beta \geq 1$.

Figure 5.4 depicts how to adapt the regret algorithm $\mathcal{R}$ to online stochastic reservation problems. The structure is similar in essence to algorithm $\mathcal{C}$. However, line 7 now credits the allocation of $r_t$ with the profit $w(\sigma^*)$ of the optimal allocation $\sigma^*$ to the scenario, while lines 8 and 9 assign some credit to the suboptimal allocations using the suboptimality approximation algorithm. Chapter 6 presents an interesting suboptimality algorithm for online stochastic reservations.

## 5.7 Cancellations

Most reservation systems allow requests to be canceled after they are accepted. It is possible to enhance the online algorithms and the sampling procedure to accommodate cancellations. It suffices to assume that an (often empty) set of cancellations $K_t$ is revealed at step $t$ in addition to the request $r_t$ and that the function SAMPLE returns pairs $\langle R, K \rangle$ of future requests $R$ and cancellations $K$. Figure 5.5 presents a revised version of the generic online algorithm: its main modification is in line 3, which removes the cancellations $K_t$ from the current assignment $\sigma_{t-1}$ before allocating a bin to the new request. Figure 5.6 depicts the consensus algorithm with cancellations, illustrating the enhanced sampling procedure (line 4) and how cancellations are taken into account when calling the optimization (line 6).

# 6 Online Multiknapsack Problems

*A deadline is negative inspiration. Still, it's better than no inspiration at all.*
— Rita Mae Brown

This chapter considers the online stochastic reservation problems studied in [6], that is, online multiknapsack problems that may allow for overbooking. It studies and compares the performance of the online stochastic algorithms on these applications. It is important to study these multiknapsack problems for three main reasons.

1. First, these applications capture a different kind of decision, since the issue is not as much to decide which requests to serve, but rather to decide how to serve selected requests. They also feature an objective function that is not the sum of the rewards of the individual requests, since, in presence of overbooking, the objective function features penalty terms assessed at the time horizon.

2. The underlying offline problems are integer-programming problems and their computational difficulty lies between packet scheduling, which can be solved in polynomial time, and multiple vehicle routing with time windows, which are extremely challenging to solve optimally.

3. Algorithms $\mathcal{E}$, $\mathcal{C}$, and $\mathcal{R}$ are compared with some other interesting stochastic algorithms. In particular, they are compared with a version of $\mathcal{E}$ where the optimization algorithm is replaced by a best-fit heuristic that is excellent in the instances considered in the experimental results. Moreover, they are also compared with an algorithm that solves, at time $t$, a single optimization problem on the expected requests coming between $t$ and the deadline.

## 6.1 Online Multiknapsack with Deadlines

**Relevance**    A noteworthy application of online reservation systems comes from travel agencies that must fill holiday centers with school groups in presence of uncertain requests and cancellations [6]. A typical instance considers reservations for a specific week. Each school reservation is characterized by the size of its group and by a price. The schools must make their reservation prior to the considered week but cannot specify a holiday center. The travel agency, on the other hand, has the ability to reject requests but, once a request is accepted, it must specify the holiday center for the request and cannot change this allocation. Schools can cancel their reservations at no cost and the travel agency can overbook the holiday centers, in which case each additional person must be accommodated in a nearby hotel at a fixed cost. The goal of the agency is to maximize its profit.

**Multiknapsack Problems**    The multiknapsack problem is an example of a reservation problem. Here each request $r$ is characterized by a reward $w(r)$ and a capacity $c(r)$. The goal is to allocate a subset $T$ of the requests $R$ to the bins $B$ so that the capacities of the bins are not exceeded and the objective function

$$w(T) = \sum_{r \in T} w(r)$$

is maximized. The multiknapsack problem is thus an instance of the generic online reservation problem where the objective function is

$$w(\sigma) = \sum_{\substack{r \in R \\ \sigma(r) \neq \perp}} w(r)$$

and the constraints are specified as follows:

$$C(\sigma) \equiv \forall b \in B : \sum_{r \in bin(b, \sigma)} c(r) \leq C_b$$

where

$$bin(b, \sigma) = \{r \in R \mid \sigma(r) = b\}.$$

The offline problem can be solved by integer programming. The integer-programming formulation associates with each request $r$ and bin $b$ a binary variable $x[r, b]$ whose value is 1 when the request is allocated to bin $b$ and 0 otherwise. The integer program can be expressed as

$$\max \quad \sum_{r \in R} \sum_{b \in B} w(r)\, x[r, b]$$

such that

$$\sum_{b \in B} x[r, b] \leq 1 \quad (r \in R)$$

$$\sum_{r \in R} c(r)\, x[r, b] \leq C_b \quad (b \in B)$$

$$x[r, b] \in \{0, 1\} \quad (r \in R, b \in B).$$

**Multiknapsack Problems with Overbooking**   In practice, many reservation systems allow for overbooking. The multiknapsack problem with overbooking allows the bin capacities to be exceeded, but overbooking is penalized in the objective function. To adapt the mathematical-programming formulation above, it suffices to introduce a nonnegative variable $y[b]$ representing the excess for each bin $b$ and to introduce a penalty term $\alpha \times y[b]$ in the objective function. The integer-programming model now becomes

$$\max \quad \sum_{r \in R} \sum_{b \in B} w(r)\, x[r,b] - \sum_{b \in B} \alpha\, y[b]$$

such that

$$\sum_{b \in B} x[r,b] \leq 1 \quad (r \in R)$$

$$\sum_{r \in R} c(r)\, x[r,b] \leq C_b + y[b] \quad (b \in B)$$

$$x[r,b] \in \{0,1\} \quad (r \in R, b \in B)$$

$$y[b] \geq 0 \quad (b \in B).$$

This is the offline problem considered in [6].

**The Online Problem**    In the online problems, the requests arrive online one at a time. Accepted requests also can be canceled, in which the capacity they take can be recovered and reused.

## 6.2   The Suboptimality Approximation

This section describes an amortized suboptimality approximation for multiknapsack problems. Given a set of requests $R$, a request $r \in R$, and an optimal solution $\sigma^*$ to the multiknapsack problem, the suboptimality algorithm must return an approximation to the optimal solution $\sigma^s$ when request $r$ is allocated to bin $b$ (that is, $\sigma^s(r) = b$) for all bin $b \in B_\perp$. The amortized suboptimality approximation must run in the time taken by a constant number of optimizations.

The key idea behind the suboptimality algorithm is to solve a number of one-dimensional knapsack problems (which take pseudo-polynomial time). There are two main cases to study:

1. Request $r$ is allocated to a bin $b$ in $B$ in solution $\sigma^*$. Then the algorithm must approximate the optimal solutions in which $r$ is allocated to $B \setminus \{b\}$ (procedure REGRET-SWAP) or rejected (procedure REGRET-SWAP-OUT).

2. Request $r$ is rejected (that is, $\sigma^*(r) = \perp$). Then the algorithm must insert request $r$ in each of the bins (procedure REGRET-SWAP-IN).

**Conventions and Notations**    Without loss of generality, we assume that the bins are numbered $1..n$ and we present three algorithms:

- REGRET-SWAP: moving request $i$ from bin 2 to bin 1;

- REGRET-REJECT: rejecting request $i$ from bin 1;

- REGRET-ACCEPT: accepting rejected request $i$ into bin 1.

REGRET-SWAP$(i)$

1   $A \leftarrow bin(1, \sigma^*) \cup bin(2, \sigma^*) \cup U(\sigma^*) \setminus \{i\}$;

2   **if** $C_1 - c(i) \geq C_2$ **then**

3      $bin(1, \sigma^a) \leftarrow knapsack(A, C_1 - c(i)) \cup \{i\}$;

4      $bin(2, \sigma^a) \leftarrow knapsack(A \setminus bin(1, \sigma^a), C_2)$;

5   **else**

6      $bin(2, \sigma^a) \leftarrow knapsack(A, C_2)$;

7      $bin(1, \sigma^a) \leftarrow knapsack(A \setminus bin(2, \sigma^a), C_1 - c(i)) \cup \{i\}$;

8   $e \leftarrow argmax(r \in bin(1, \sigma^*) \setminus bin(1..2, \sigma^a) : c(r) > \max(C_1 - c(i), C_2))\ w(r)$;

9   **if** $e$ exists & $w(e) > \max(w(bin(1, \sigma^a)), w(bin(2, \sigma^a)))$ **then**

10     $j \leftarrow argmax(j \in 3..n)\ C_j$;

11     $bin(j, \sigma^a) \leftarrow knapsack(bin(j, \sigma^a) \cup \{e\}, C_j)$;

**Figure 6.1:** The Suboptimality Algorithm for the Knapsack Problem: Swapping $i$ from Bin 2 to Bin 1.

In the following, $\sigma^*$ represents the optimal solution to the multiknapsack problem, $\sigma^s$ denotes the optimal solution in which request $i$ is moved, and $\sigma^a$ is the suboptimality approximation. The solution to the one-dimensional knapsack problem on $R$ for a bin with capacity $C$ is denoted by $knapsack(R, C)$. We use $c(R)$ to denote the sum of the capacities of the requests in $R$, that is,

$$c(R) = \sum_{r \in R} c(r),$$

and $U(\sigma^*)$ to denote the rejected requests in $\sigma^*$

$$U(\sigma^*) = \{r \in \sigma^* \mid \sigma^*(r) = \bot\}.$$

Finally, we lift the notation $bin(b, \sigma)$ from bins to sets of bins.

## 6.2.1 Swapping a Request between Two Bins

Figure 6.1 depicts the algorithm to swap request $i$ from bin 1 to bin 2. The key idea is to consider all requests allocated to bins 1 and 2 in $\sigma^*$ and to solve two one-dimensional knapsack problems for bin 1 (without the capacity taken by request $i$) and bin 2. The algorithm always starts with the bin whose remaining capacity is largest. After solving these two one-dimensional knapsacks, if there exists a request $e \in bin(1, \sigma^*)$ that is not allocated in $bin(1..2, \sigma^a)$ and whose value is higher than the values of these two bins, the algorithm solves a third knapsack problem to place this request in another bin if appropriate. This is important if request $e$ is of high value but cannot be allocated in bin 1 due to the capacity taken by request $i$.

THEOREM 6.1   Algorithm REGRET-SWAP is a constant-factor approximation.

PROOF:     Let $\sigma^s$ be the suboptimal solution, $\sigma^a$ be the regret solution, and $\sigma^*$ be the optimal solution. Consider the following sets

$$
\begin{aligned}
I_1 &= bin(B, \sigma^s) \cap bin(B, \sigma^a) \\
I_2 &= (bin(1, \sigma^s) \setminus bin(B, \sigma^a)) \cap U(\sigma^*) \\
I_3 &= (bin(2, \sigma^s) \setminus bin(B, \sigma^a)) \cap U(\sigma^*) \\
I_4 &= (bin(3..n, \sigma^s) \setminus bin(B, \sigma^a)) \cap U(\sigma^*) \\
I_5 &= (bin(1, \sigma^s) \setminus bin(B, \sigma^a)) \cap bin(1, \sigma^*) \\
I_6 &= (bin(1, \sigma^s) \setminus bin(B, \sigma^a)) \cap bin(2, \sigma^*).
\end{aligned}
\qquad
\begin{aligned}
I_7 &= (bin(2, \sigma^s) \setminus \sigma^a) \cap bin(1, \sigma^*) \\
I_8 &= (bin(2, \sigma^s) \setminus bin(B, \sigma^a)) \cap bin(2, \sigma^*) \\
I_9 &= (bin(3..n, \sigma^s) \setminus bin(B, \sigma^a)) \cap bin(1, \sigma^*) \\
I_{10} &= (bin(3..n, \sigma^s) \setminus bin(B, \sigma^a)) \cap bin(2, \sigma^*) \\
I_{11} &= (bin(1..n, \sigma^s) \setminus bin(B, \sigma^a)) \cap bin(3..n, \sigma^*)
\end{aligned}
$$

The suboptimal solution $\sigma^s$ can be partitioned into

$$\sigma^s = \bigcup_{k=1}^{11} I_k$$

and the proof shows that

$$w(I_k) \le \beta_k \, w(\sigma^a) \quad (1 \le k \le 13)$$

which implies that

$$w(\sigma^s) \le \beta \, w(\sigma^a)$$

for some constant $\beta = \beta_1 + \ldots \beta_{11}$. The proof of each inequality typically separates into two cases:

**A:** $C_1 - c(i) \ge C_2$;

**B:** $C_1 - c(i) < C_2$.

Case A executes lines 3 and 4 and case B executes line 6 and 7. Observe also that the proof that $w(I_1) \le w(\sigma^a)$ is immediate. We now give the proofs for the remaining sets. In the proofs, $C_1'$ denotes $C_1 - c(i)$ and $K(E, C)$ is defined as follows:

$$K(E, C) = w(knapsack(E, C)).$$

$I_2.A$ :  $K(I_2, C_1') \le K(U(\sigma^*), C_1') \le K(bin(1, \sigma^a), C_1') \le w(\sigma^a).$

$I_2.B$ :  $K(I_2, C_1') \le K(U(\sigma^*), C_1') \le K(U(\sigma^*), C_2) \le K(bin(2, \sigma^a), C_2) \le w(\sigma^a).$

$I_3.A$ :  $K(I_3, C_2) \le K(U(\sigma^*), C_2) \le K(U(\sigma^*), C_1') \le K(bin(1, \sigma^a), C_1') \le w(\sigma^a).$

$I_3.B$ :  $K(I_3, C_2) \le K(U(\sigma^*), C_2) \le K(bin(2, \sigma^a), C_2) \le w(\sigma^a).$

$I_4$ : Assume that $w(I_4) > w(\sigma^a)$. This implies

$$w(I_4) \quad > \quad w(bin(1, \sigma^a)) + w(bin(2, \sigma^a)) + w(bin(3..n, \sigma^a)) \geq w(bin(3..n, \sigma^a)) \geq w(bin(3..n, \sigma^*)),$$

which contradicts the optimality of $\sigma^*$ since $I_4 \subseteq U(\sigma^*)$.

$I_5.A$ : $K(I_5, C_1') \leq K(bin(1, \sigma^*), C_1') \leq K(A, C_1') \leq w(bin(1, \sigma^a)) \leq w(\sigma^a)$.

$I_5.B$ : $K(I_5, C_1') \leq K(bin(1, \sigma^*), C_2) \leq K(A, C_2) \leq K(bin(2, \sigma^a), C_2) \leq w(\sigma^a)$.

$I_6.A$ : $K(I_6, C_1') \leq K(bin(2, \sigma^*) \setminus \{i\}, C_1') \leq K(bin(1, \sigma^a), C_1') \leq w(\sigma^a)$

$I_6.B$ : $K(I_6, C_1') \leq K(bin(2, \sigma^*) \setminus \{i\}, C_2) \leq K(bin(2, \sigma^a), C_2) \leq w(\sigma^a)$

$I_7.A$ : $K(I_7, C_2) \leq K(I_7, C_1') \leq K(bin(1, \sigma^*), C_1') \leq K(bin(1, \sigma^a), C_1') \leq w(\sigma^a)$.

$I_7.B$ : $K(I_7, C_2) \leq K(bin(1, \sigma^*), C_2) \leq K(bin(2, \sigma^a), C_2) \leq w(\sigma^a)$.

$I_8.A$ : $K(I_8, C_2) \leq K(I_8, C_1') \leq K(bin(2, \sigma^*), C_1') \leq K(bin(1, \sigma^a), C_1') \leq w(\sigma^a)$

$I_8.B$ : $K(I_8, C_2) \leq K(bin(2, \sigma^*), C_2) \leq K(bin(2, \sigma^a), C_2) \leq w(\sigma^a)$.

$I_9.A$ : Consider

$$T \quad = \quad knapsack(bin(1, \sigma^*), C_1');$$
$$L \quad = \quad bin(1, \sigma^*) \setminus T$$

and let $e = \text{arg-max}_{e \in L}\, w(e)$. By optimality of $T$, we know that $c(T) + c(e) > C_1'$ and, since $bin(1, \sigma^*) = T \cup L$, we have that $c(L \setminus \{e\}) < c(i)$.

If $w(e) \leq \max(w(bin(1, \sigma^a)), w(bin(2, \sigma^a)))$, then

$$
\begin{aligned}
w(I_9) \quad &\leq \quad w(T) + w(L \setminus \{e\}) + w(e) \\
&\leq \quad w(bin(1, \sigma^a)) + w(bin(2, \sigma^a)) + w(e) \\
&\leq \quad 2(w(bin(1, \sigma^a)) + w(bin(2, \sigma^a))) \leq 2w(\sigma^a).
\end{aligned}
$$

Otherwise, by optimality of $bin(1, \sigma^a)$ and $bin(2, \sigma^a)$, we have that

$$c(e) > C_1' \,\&\, c(e) > C_2$$

and the algorithm executes lines 10 and 11. If $c(e) \leq C_j$, then

$$
\begin{aligned}
w(I_9) \quad &\leq \quad w(T) + w(L \setminus \{e\}) + w(e) \\
&\leq \quad w(bin(1, \sigma^a)) + w(bin(2, \sigma^a)) + w(bin(j, \sigma^a)) \leq w(\sigma^a).
\end{aligned}
$$

Otherwise, if $c(e) > C_j$, $e \notin \sigma^s$ and

$$w(I_9) \quad \leq \quad w(T) + w(L \setminus \{e\}) \leq w(bin(1, \sigma^a)) + w(bin(2, \sigma^a)) \leq w(\sigma^a).$$

$I_9.B$ : Consider

$$T = knapsack(bin(1, \sigma^*), C_2);$$
$$L = bin(1, \sigma^*) \setminus T$$

and let $e = \text{arg-max}_{e \in L} w(e)$. If $w(T) \geq w(L)$, we have that

$$w(bin(1, \sigma^*)) \leq 2w(T) \leq 2w(bin(2, \sigma^a)) \leq 2w(\sigma^a).$$

Otherwise, $c(L) > C_2$ by optimality of $T$ and thus $c(L) > c(i)$ since $C_2 \geq c(i)$. By optimality of $T$, $c(T \cup \{e\}) > C_2 > C_1'$ and, since $bin(1, \sigma^*) = T \cup L$, it follows that $c(L \setminus \{e\}) \leq c(i)$ Hence $w(L \setminus \{e\}) \leq w(T)$ by optimality of $T$ and

$$w(I_9) \quad \leq \quad w(T) + w(L \setminus \{e\}) + w(e) \leq 2w(T) + w(e) \leq 2w(bin(2, \sigma^a)) + w(e).$$

If $w(e) \leq w(bin(2, \sigma^a))$, $w(I_9) \leq 3w(bin(2, \sigma^a)) \leq 3w(\sigma^a)$ and the result follows. Otherwise, by optimality of $bin(2, \sigma^a)$, $c(e) > C_2 \geq C_1'$ and the algorithm executes lines 10 and 11. If $c(e) \leq C_j$, then

$$w(I_9) \quad \leq \quad 2w(bin(1, \sigma^a)) + w(bin(j, \sigma^a)) \leq w(\sigma^a).$$

Otherwise, if $c(e) > C_j$, $e \notin \sigma^s$ and

$$w(I_9) \quad \leq \quad w(T) + w(L \setminus \{e\}) \leq 2w(bin(2, \sigma^a)) \leq 2w(\sigma^a).$$

$I_{10}.A$ : $w(I_{10}) \leq w(bin(2, \sigma^*)) - w(i) \leq w(bin(1, \sigma^a)) \leq w(\sigma^a)$.

$I_{10}.B$ : $w(I_{10}) \leq w(bin(2, \sigma^*)) - w(i) \leq w(bin(2, \sigma^a)) \leq w(\sigma^a)$.

$I_{11}$ : $K(bin(3..n, \sigma^*)) \leq K(3..n, \sigma^a)$. $\qquad\qquad\qquad\qquad\qquad\qquad\qquad\qquad$ $\square$

## 6.2.2   Swapping a Request out of a Bin

The algorithm to swap a request $i$ out of bin 1 is depicted in figure 6.2. It consists of solving a one-dimensional knapsack with the rejected requests and the requests already in that bin. The proof is similar, but simpler, to the proof of theorem 6.1.

THEOREM 6.2   Algorithm REGRET-SWAP-OUT is a constant-factor approximation.

PROOF:    Let $\sigma^s$ be the suboptimal solution, $\sigma^r$ be the regret solution, and $\sigma^*$ be the optimal

REGRET-REJECT$(i)$

1    $A \leftarrow bin(1, \sigma^*) \cup U(\sigma^*) \setminus \{i\}$;

2    $bin(1, \sigma^a) \leftarrow knapsack(A, C_1)$;

**Figure 6.2:** The Suboptimality Algorithm for the Knapsack Problem: Swapping $i$ out of Bin 1.

solution. Consider the following sets

$$
\begin{aligned}
I_1 &= bin(B, \sigma^s) \cap bin(B, \sigma^r) \\
I_2 &= (bin(1, \sigma^s) \setminus bin(\sigma^r)) \cap U(\sigma^*) \\
I_3 &= (bin(1, \sigma^s) \setminus bin(\sigma^r)) \cap bin(1, \sigma^*) \\
I_4 &= (bin(2..n, \sigma^s) \setminus bin(\sigma^r)) \cap U(\sigma^*) \\
I_5 &= (bin(2..n, \sigma^s) \setminus bin(\sigma^r)) \cap bin(1, \sigma^*) \\
I_6 &= (bin(1..n, \sigma^s) \setminus bin(\sigma^r)) \cap bin(2..n, \sigma^*)
\end{aligned}
$$

The suboptimal solution $\sigma^s$ can be partitioned into

$$
\sigma^s = \bigcup_{k=1}^{6} I_k
$$

and the proof shows that

$$
w(I_k) \leq \beta_k \, w(\sigma^r) \quad (1 \leq k \leq 6),
$$

which implies that

$$
w(\sigma^s) \leq \beta \, w(\sigma^r)
$$

for some constant $\beta = \beta_1 + \ldots \beta_7$. The proof is immediate for $I_1$. Since $K(bin(2..n, \sigma^*)) \leq K(2..n, \sigma^r)$, the result also follows for $I_6$.

$I_2$ : By lines 1 and 2 of the algorithm, $K(I_2, C_1) \leq K(U(\sigma^*), C_1) \leq K(bin(1, \sigma^r), C_1) \leq w(\sigma^r)$.

$I_3$ : By lines 1 and 2 of the algorithm, $K(I_3, C_1) \leq K(bin(1, \sigma^*) \setminus \{i\}, C_1) \leq K(bin(1, \sigma^r), C_1) \leq w(\sigma^r)$.

$I_4$ : By optimality of $O$, $w(I_4) \leq w(bin(2..n, \sigma^*)) = w(bin(2..n, \sigma^r)) \leq w(\sigma^r)$.

$I_5$ : Since $c(bin(1, \sigma^*)) \leq C_1$, $w(I_5) \leq w(bin(1, \sigma^*) \setminus \{i\}) \leq w(bin(1, \sigma^r)) \leq w(\sigma^r)$.

□

REGRET-SWAP-IN$(i, 1)$

1   $A \leftarrow bin(1, \sigma^*) \cup U(\sigma^*)$;

2   $bin(1, \sigma^a) \leftarrow knapsack(A, C_1 - c(i)) \cup \{i\}$;

3   $L \leftarrow bin(1, \sigma^*) \setminus bin(1, \sigma^a)$;

4   **if** $w(L) > w(bin(1, \sigma^a))$ **then**

5      $j \leftarrow argmax(j \in 2..n) \; C_j$;

6      $bin(j, \sigma^a) \leftarrow knapsack(bin(j, \sigma^a) \cup L, C_j)$;

**Figure 6.3:** The Suboptimality Algorithm for the Knapsack Problem: Swapping $i$ into Bin 1.

## 6.2.3  Swapping a Request into a Bin

Figure 6.3 depicts the algorithm for swapping a request $i$ in bin 1, which is essentially similar to REGRET-SWAP but uses only one bin. It assumes that request $i$ can be placed in at least two bins, since otherwise a single additional optimization suffices to compute the suboptimality approximations for all bins $b \in B$. Once again, it solves a one-dimensional knapsack for bin 1 (after having allocated request $i$) with all the requests in $bin(1, \sigma^*)$ and the unallocated requests. If the resulting knapsack is of low quality (that is, the remaining requests from $bin(1, \sigma^*)$ have a higher value than $bin(1, \sigma^a)$), REGRET-SWAP-IN solves an additional knapsack problem for the largest available bin. The proof is once again similar to the proof of theorem 6.1.

THEOREM 6.3  Assuming that item $i$ can be placed in at least two bins, Algorithm REGRET-SWAP-IN is a constant-factor approximation.

PROOF:  Let $\sigma^s$ be the suboptimal solution, $\sigma^r$ be the regret solution, and $\sigma^*$ be the optimal solution. Consider the following sets

$$
\begin{aligned}
I_1 &= bin(B, \sigma^s) \cap bin(B, \sigma^r) \\
I_2 &= (bin(1, \sigma^s) \setminus \sigma^r) \cap U(\sigma^*) \\
I_3 &= (bin(2..n, \sigma^s) \setminus \sigma^r) \cap U(\sigma^*) \\
I_4 &= (bin(1, \sigma^s) \setminus \sigma^r) \cap bin(1, \sigma^*) \\
I_5 &= (bin(2..n, \sigma^s) \setminus \sigma^r) \cap bin(1, \sigma^*) \\
I_6 &= (bin(1..n, \sigma^s) \setminus \sigma^r) \cap bin(2..n, \sigma^*).
\end{aligned}
$$

The suboptimal solution $\sigma^s$ can be partitioned into

$$\sigma^s = \bigcup_{k=1}^{6} I_k$$

and the proof shows that

$$w(I_k) \leq \beta_k \, w(\sigma^r) \quad (1 \leq k \leq 6),$$

which implies that

$$w(\sigma^s) \leq \beta \, w(\sigma^r)$$

for some constant $\beta = \beta_1 + \ldots \beta_6$. Observe also that the proof that $w(I_1) \leq w(\sigma^r)$ is immediate. We now give the proofs for the remaining sets. In the proofs, $C_1'$ denotes $C_1 - c(i)$ and $K(E, C)$ is defined as follows:

$$K(E, C) = w(knapsack(E, C)).$$

$I_2 \ : \ K(I_2, C_1') \leq K(U(\sigma^*), C_1') \leq K(bin(1, \sigma^r), C_1') \leq w(\sigma^r).$

$I_3 \ : \ $ Assume that $w(I_3) > w(\sigma^r)$. This implies

$$w(I_3) > w(bin(1, \sigma^r)) + w(bin(2..n, \sigma^r)) \geq w(bin(2..n, \sigma^r)) \geq w(bin(2..n, \sigma^*)),$$

which contradicts the optimality of $\sigma^*$ since $I_3 \subseteq U(\sigma^*)$.

$I_4. \ : \ K(I_4, C_1') \leq K(bin(1, \sigma^*), C_1') \leq K(A, C_1') \leq w(bin(1, \sigma^r)) \leq w(\sigma^r).$

$I_5 \ : \ $ Consider

$$
\begin{aligned}
T &= knapsack(bin(1, \sigma^*), C_1'); \\
L &= bin(1, \sigma^*) \setminus T
\end{aligned}
$$

and let

$$e = \operatorname*{arg\text{-}max}_{e \in L} w(e).$$

By optimality of $T$, we know that

$$c(T) + c(e) > C_1'$$

and, since $bin(1, \sigma^*) = T \cup L$, we have that

$$c(L \setminus \{e\}) < c(i).$$

If $w(L) \leq w(bin(1, \sigma^r))$ then,

$$
\begin{aligned}
w(I_5) &\leq w(T) + w(L) \\
&\leq 2w(\sigma^r).
\end{aligned}
$$

Otherwise, the algorithm executes lines 5 and 6. By assumption, $C_j \geq c(i) \geq c(L \setminus \{e\})$ and hence $w(L) \leq w(bin(j, \sigma^r))$ by optimality of the knapsack in line 6. If $c(e) \leq C_j$, then

$$
\begin{aligned}
w(I_5) &\leq w(T) + w(L \setminus \{e\}) + w(e) \\
&\leq w(bin(1, \sigma^r)) + w(bin(j, \sigma^r)) + w(bin(j, \sigma^r)) \\
&\leq 2w(\sigma^r).
\end{aligned}
$$

Otherwise, if $c(e) > C_j$, $e \notin \sigma^s$ and

$$
\begin{aligned}
w(I_5) &\leq w(T) + w(L \setminus \{e\}) \\
&\leq w(bin(1, \sigma^r)) + w(bin(j, \sigma^r)) \\
&\leq w(\sigma^r).
\end{aligned}
$$

$I_6 \ : \ K(bin(2..n, \sigma^*)) \leq K(2..n, \sigma^r).$

$\square$

## 6.3   Experimental Results

This section presents the experimental results on multiknapsack problems. Section 6.3.1 describes the instances considered in this chapter and section 6.3.2 describes some alternative algorithms that are used for contrasting the online stochastic algorithms. Section 6.3.3 compares the main algorithms. Section 6.3.4 studies the anticipativity assumption in this setting. Section 6.3.5 studies the tradeoff between the number of samples and the quality of the underlying optimization algorithm. Section 6.3.6 then demonstrates the benefits of sampling.

### 6.3.1   The Instances

The experimental results are based on benchmarks essentially similar to those in [6], but use the more standard Poisson arrival processes for the requests. Requests are classified in $k$ types, each type being characterized by a capacity and a value. The arrivals of requests of each type $i$ follow a Poisson process with parameter $\lambda_i$. The cancellations are generated from a distribution of the time spent in the system. For a request $r$ of type $i$, the total time request $r$ spends in the system follows an exponential distribution with parameter $\theta_i$. As a consequence, a request $r$ is canceled if its departure from the system is before the time horizon $h$ and is not canceled otherwise. The arrival processes for each request type are independent. The cancellations of the requests are independent of each other as well. More precisely, the relevant probabilities are specified as follows:

$$A_i(t) = \Pr[\text{next request of type } i \text{ arrives in the next } t \text{ time steps}] = 1 - e^{-\lambda_i t}$$
$$K_i(t) = \Pr[\text{existing request } r \text{ of type } i \text{ departs in the next } t \text{ steps}] = 1 - e^{-\theta_i t}.$$

As in [6], the instances are generated from a master problem with the following features: $|B| = 5$ bins, each with capacity of 100 units of weight, and $k = 5$ different types of items, with an average weight and an average value equal to 25. The weights of the 5 types are $\{17, 20, 25, 30, 33\}$, the values are $\{13, 26, 21, 26, 39\}$ and the overbooking penalty $\alpha$ is 10. The arrival rate $\lambda_i$ is 0.008 and the parameter $\theta_i = (\ln 2)/1000 = 0.000693$ for all request type $i$. The time horizon in the instance is $h = 1000$. As a result, the expected capacity of the arriving items is twice the total capacity of the bins, since there are 8 expected requests of each type with average capacity of 25 for an expected total capacity of $8 \times 25 \times 1000 = 2 \cdot |B| \cdot 100$ unit. Note that the value chosen for parameter $\theta_i$ implies that requests arriving at the start of the online algorithm have a probability $1/2$ of subsequently being canceled. The total capacity of the arriving requests that are not canceled is thus around 145 percent of the total bin capacity.

The set of instances, denoted by A through I, were generated from the master problem. The goal was to produce a diverse set of problems revealing the strengths and weaknesses of the various algorithms. Problem A scales the master problem by doubling the weight and value of the request types in the master problem, as well as halving the number of items that arrive. Problem B further scales problem A by increasing the weight and value of the types. Problem C considers seven types of items whose cost ratio takes the form of a bell shape. Problem D looks at the master problem and doubles the number of bins while dividing their capacity by two. Problem E considers a version of the master problem with bins of variable capacity. Problem F depicts a version of the master problem whose items arrive three times as often and cancel three times as often. Problem G considers a much larger problem with thirty five request types whose cost ratio is also shaped in a bell. Problem H is like problem G; the main difference is that the cost ratio shape is reversed. Problem I is a version of G with an extra bin.

### 6.3.2 Alternative Algorithms

Several other algorithms are used for comparison on online stochastic knapsacks.

**The Greedy Best-Fit Heuristic** The greedy best-fit (BF) algorithm consists of accepting the requests greedily and using a best-fit strategy to select the bin to serve the request. The best-fit heuristic, which is illustrated in figure 6.4, can be specified as

CHOOSEALLOCATION-G$(\sigma_{t-1}, r_t)$
1    **return** $argmin(b \in \mathcal{F}(\sigma_{t-1}, r_t) \setminus \{B_\perp\})\ C_b(\sigma_{t-1})$;

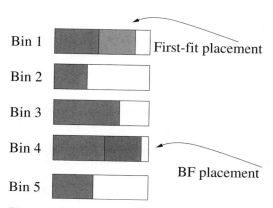

Bin 1    First-fit placement

Bin 2

Bin 3

Bin 4    BF placement

Bin 5

**Figure 6.4:** The Best-Fit Strategy in the Greedy Algorithm.

where $C_b(\sigma)$ denotes the remaining capacity of the bin $b \in B_\perp$ in $\sigma$, that is,

$$C_b(\sigma) = C_b - c(bin(b, \sigma)).$$

**The Expected-Problem Algorithm**    An approach commonly used in presence of uncertainty is to select the allocation based on the solution to an "expected" problem. For the multiknapsack problems considered here, this problem can be obtained by generating, for each request type, the expected arrivals and cancellations since the distributions are known. It is thus important to study the benefits of sampling with respect to an algorithm that bases its decision on an expected problem.

For the instances considered in this chapter, the expected problem can be obtained as follows: Let $X_i(t)$ be the random variable denoting the total number of requests of type $i$ that have arrived at time $t$ and have not been canceled. Variable $X_i(t)$ follows a Poisson distribution with mean

$$\lambda_i t p_i = \frac{\lambda_i}{\theta_i} \left(1 - e^{-\theta_i t}\right)$$

where

$$p_i = \frac{1 - e^{-\theta_i t}}{\theta_i t}$$

and

$$\Pr[X_i(t) = j] = e^{-\lambda_i t p_i} \frac{(\lambda_i t p_i)^j}{j!}.$$

See, for instance, [92] for these results. As a consequence, the expected number of new requests of type $i$ from time $t$ to the horizon $h$ is given by

$$\lambda_i(h-t)p_i = \frac{\lambda_i}{\theta_i}\left(1 - e^{-\theta_i(h-t)}\right).$$

Once these new requests are available, the expected problem is obtained by computing the expected cancellations of the available requests. Algorithm EP can then be specified as follows: it first generates the expected problem and solves it using the allocation of past requests. It then uses this optimal solution to determine the allocation of request $r$. If no request of type $r$ appears in the optimal solution, the request $r$ is rejected. Otherwise, the algorithm considers the bins where $r$ appears and selects a bin where a request of type $r$ is allocated (breaking ties randomly). Intuitively, algorithm EP can be viewed as replacing the sampling process in algorithm $\mathcal{E}$ by a single expected problem and gives us some indication about the benefits of sampling.

**The Heuristic Expectation Algorithm**   The last algorithm considered for comparison purposes is a variation of algorithm $\mathcal{E}$ where the call to the optimization algorithm $\mathcal{O}$ is replaced by an offline greedy heuristic. The algorithm, denoted by $\mathcal{E}(BF)$, applies a best-fit heuristic on each scenario after having sorted the requests in decreasing order of their value/capacity ratio, that is,

$$\frac{v(r_{i_1})}{c(r_{i_1})} \geq \ldots \geq \frac{v(r_{i_t})}{c(r_{i_t})}.$$

On the offline master problem and its variations, the best-fit heuristic performs quite well. It is 5 percent off the optimum in the average and is never worse than 10 percent off. Moreover, this algorithm is extremely fast compared to the integer-programming solver and therefore algorithm $\mathcal{E}(BF)$ is able to approximate a very large number of scenarios compared to $\mathcal{E}$, $\mathcal{R}$, and $\mathcal{C}$. As a result, it provides an helpful case study for understanding the tradeoff between the quality of the optimization algorithm and the number of scenarios. Such a tradeoff is also studied in chapter 9.

### 6.3.3   Experimental Comparisons of the Main Algorithms

This section compares the main algorithms: the greedy algorithm (G) and the online stochastic optimization algorithms $\mathcal{E}$, $\mathcal{C}$, and $\mathcal{R}$. The integer programs are solved with CPLEX 9.0 with a time limit of 10 seconds. The optimal solutions can be found within the time limit for all instances but G, H, and I. Every instance is executed under various time constraints, that is, $\mathcal{T} = 1, 5, 10, 25, 50, 100$, and the results are the average of 50 executions.

**Multiknapsack with Overbooking**   Figure 6.5 summarizes the experimental results of the main algorithm on the instances A through I. It reports the average profit and the average loss of the

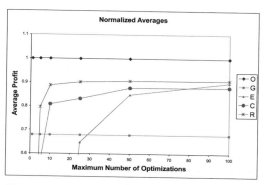
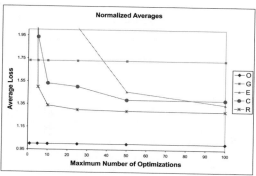

**Figure 6.5:** Experimental Results of the Main Algorithms with Overbooking: Summary.

main online algorithms as a function of the number of optimizations. The loss sums the weights of the rejected requests and the overbooking penalty (if any) and is often used in comparing online algorithms as it gives the "price" of the uncertainty.

The results show clearly the value of stochastic information as algorithms $\mathcal{R}$, $\mathcal{C}$, and $\mathcal{E}$ recover most of the gap between the online best-fit heuristic (G) and the offline optimum (which typically cannot be achieved in an online setting). Moreover, they show that algorithms $\mathcal{R}$ and $\mathcal{C}$ achieve excellent results even with small numbers of optimizations (tight time constraints). In particular, algorithm $\mathcal{R}$ achieves about 89 percent of the offline optimum with only 10 samples and over 90 percent with 25 optimizations. It also achieves a loss of 30 percent over the offline optimum for 25 optimizations and 34 percent for 10 optimizations. Algorithm $\mathcal{R}$ dominates algorithm $\mathcal{E}$, which performs poorly for tight time constraints. It becomes reasonable only for 50 optimizations and typically reaches the quality of the regret algorithm for 100 optimizations. The benefits over the greedy algorithm are quite substantial as algorithm G achieves only 68 percent of the offline optimum.

Figure 6.6 depicts the results on some selected instances: A, D, G, and H. The results show that there is considerable variation across the instances. On instance D, algorithm $\mathcal{E}$ needs a significant number of optimizations to reach some reasonable quality, while it performs much better on instance G. This instance is in fact more challenging for $\mathcal{R}$, which is dominated by $\mathcal{C}$. Note also the gap between the offline and online solutions that also vary from instance to instance.

**Multiknapsack with No Overbooking**    Figure 6.7 depicts the same results when no overbooking is allowed. These instances are easier in the sense that fewer optimizations are necessary for the algorithms to converge. But they exhibit the same pattern as when overbooking is allowed. These results show that the benefits of the regret algorithm increase with the problem complexity but are significant even on easier instances.

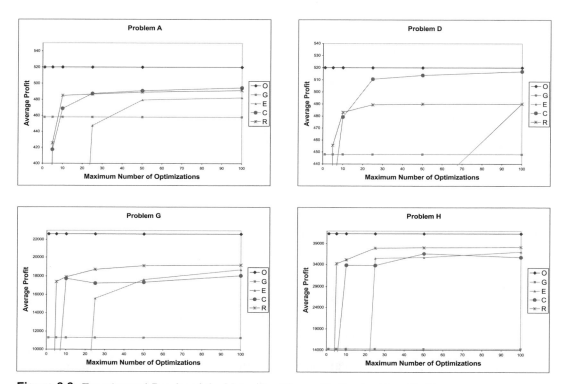

**Figure 6.6:** Experimental Results of the Main Algorithms with Overbooking: Profits.

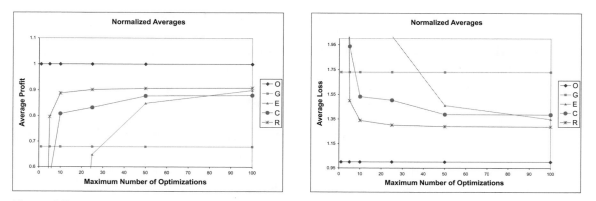

**Figure 6.7:** Experimental Results of the Main Algorithms with No Overbooking: Summary.

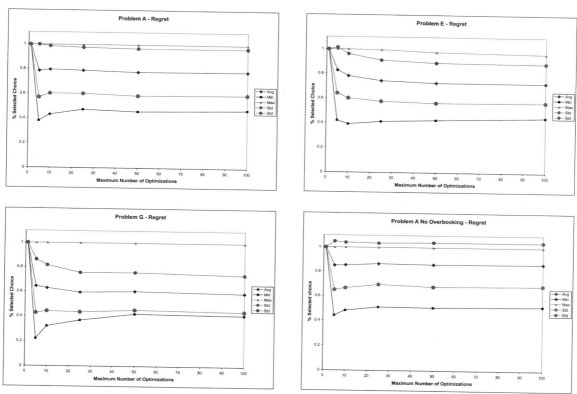

**Figure 6.8:** The Anticipativity Assumption in Multiknapsack Problems.

## 6.3.4 The Anticipativity Assumption

Figure 6.8 studies the anticipativity assumption for online stochastic multiknapsacks. It depicts how many times (in percentage) the allocation chosen in algorithm $\mathcal{R}$ is also selected in the various scenarios. More precisely, for each incoming request, the algorithm computed how many times the selected allocation was chosen in the scenarios, and the figures report the aggregate results for all times. The figure reports the average, minimum, and maximum values (in percentage) as well as the average plus or minus the standard deviation. Only representative instances, that is, A, E, and G with overbooking are shown, as well as G without overbooking.

For the overbooking instances, the results show that the selected reservation is indeed the choice in about 78 percent, 72 percent, and 58 percent of the scenarios in the average for the instances $A$, $E$, and $G$. For the instances without overbooking, the selected reservation is also the scenario choice in

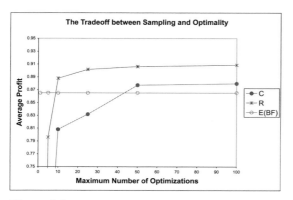
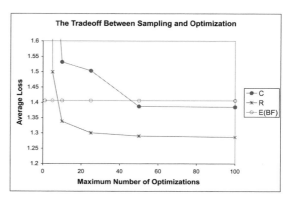

**Figure 6.9:** The Tradeoff between Sampling and Optimization: E(BF) versus $\mathcal{R}$ and $\mathcal{C}$ (Summary).

about 86 percent, 73 percent, and 60 percent of the scenarios. Although the multiknapsack problems are more challenging than packet scheduling, these results indicate once again that the anticipativity assumption is valid to a large extent and justifies the compromise between computational time and evaluation accuracy advocated in this book.

### 6.3.5 The Tradeoff between Sampling and Optimization

Figure 6.9 aggregates the experimental results comparing $\mathcal{R}$ and $\mathcal{E}(\text{BF})$, while figure 6.10 depicts the results for some selected instances. Recall that algorithm $\mathcal{E}(\text{BF})$ replaces the optimization algorithm $\mathcal{O}$ by the best-fit heuristic inside algorithm $\mathcal{E}$. Since the best-fit heuristic is extremely fast compared to the integer-programming solver, algorithm $\mathcal{E}(\text{BF})$ can approximate ten thousand scenarios within the time taken by an optimization. Moreover, generating even more scenarios does not bring any additional benefits in terms of the quality of the decisions. This explains why, in the figures, the results for algorithm $\mathcal{E}(\text{BF})$ are always straight lines. The results show that $\mathcal{E}(\text{BF})$ indeed produces excellent results but is quickly dominated by $\mathcal{R}$ as time increases. In particular, algorithm $\mathcal{E}(\text{BF})$ is almost always dominated by the regret algorithm with 10 optimizations, which corresponds to less than 2 minutes in the experimental setting. It is for instance H only that regret needs 25 optimizations. In the average, the loss of $\mathcal{E}(\text{BF})$ is 41 percent, while the regret algorithm has a loss of 34 percent for 10 optimizations and 28 percent for 25 optimizations. Similarly, the profit increases by 2 percent with 10 optimizations and 4 percent with 25 optimizations.

What is quite remarkable here is that the 5 percent difference in quality between the best-fit heuristic and the offline algorithm translates into a similar difference in quality in the online setting. Moreover, when looking at specific instances, one can see that $\mathcal{E}(\text{BF})$ is often comparable to $\mathcal{R}$, but its loss (respectively profit) may be significantly higher (respectively lower) on more difficult

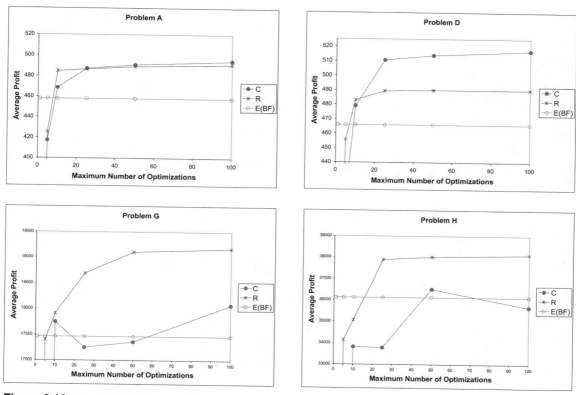

**Figure 6.10:** The Tradeoff between Sampling and Optimization: $E(BF)$ versus $\mathcal{R}$ and $\mathcal{C}$.

instances. This is the case for instances D, G, and H, in which the gap between the offline solutions and the solutions by algorithm $\mathcal{R}$ is larger. This seems to indicate that the harder the problems the more beneficial algorithm $\mathcal{R}$ becomes. We will come back to this issue for vehicle dispatching.

## 6.3.6 The Importance of Sampling

Figure 6.11 depicts the results of algorithm EP on the overbooking instances and compares it with the greedy and regret algorithms. Not all instances are shown since algorithm EP fails on some of them, returning a huge penalty because it was expecting many more cancellations. However, even on other instances, algorithm EP performs poorly. It thus seems critical to sample the future on these applications in order to take the proper decisions.

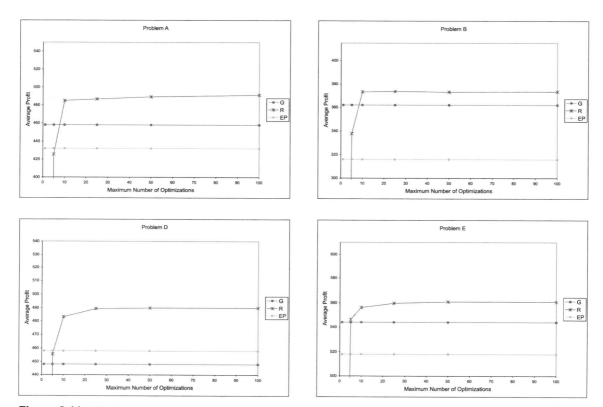

**Figure 6.11:** The Importance of Sampling: EP versus $\mathcal{R}$ and $\mathcal{C}$.

## 6.4  Notes and Further Reading

This chapter was motivated by reference [6], which proposed the problem and its basic instances and compared a variety of algorithms such as the greedy algorithm, EP, $\mathcal{E}(\mathrm{BF})$, as well as yield management and approximated Markov decision processes. In their experiments, $\mathcal{E}(\mathrm{BF})$ was shown to be superior to all other approaches on essentially all the instances. The MDP approach did not scale to the full problem and was used only for the last stage of the problems.

The results described in this chapter were first reported in [113]. The goal was to investigate whether more sophisticated optimization algorithms, together with the computational benefits of the consensus and regret algorithms, would improve upon $\mathcal{E}(\mathrm{BF})$. Since the best-fit algorithm is only 5 percent below the optimal offline solution on average, the improvements produced by the regret algorithm gave valuable information on the tradeoff between the number of scenarios and the accuracy of the optimization algorithms. Indeed earlier results on vehicle dispatching had shown that it was better to use a sophisticated heuristic on fewer scenarios, but the simpler heuristic was dominated in quality. On the multiknapsack problems, the heuristic is of high quality, yet a more sophisticated optimization algorithm, together with the regret algorithm, produces better results as soon as 10 or more optimizations are available. The results are also revisited in [77] where the algorithms exploit offline integer programs capturing cancellations directly.

# III ONLINE STOCHASTIC ROUTING

# 7 Vehicle Routing with Time Windows

*One out of four people is this country is mentally unbalanced. Think of your three closest friends and if they seem OK then you're the one.*
— Ann Landers

This chapter reviews the vehicle dispatching and routing problems whose online versions are studied in the next three chapters. It reviews the main concepts in this area and presents an effective algorithm to find high-quality solutions to these problems.

Vehicle routing and dispatching problems are important components of many distribution and transportation systems including bank deliveries, postal deliveries, school bus routing, and security patrol services. Many applications, like courier services, taxi services, repair services, and package pickup and delivery have inherent uncertainty associated with their customers (that is, customers make requests during solution execution).

It is important to emphasize that vehicle routing and dispatching are combinatorial optimization problems that are extremely hard to solve. For instance, many Solomon benchmarks involve about one hundred customers and have not been solved optimally at the time of this writing despite considerable research. Even more interesting perhaps, the best-known solutions have improved steadily over the past decades, producing increasingly sophisticated and innovative algorithms. As a consequence, these problems arising in numerous industrial applications present significant challenges for online stochastic optimization.

## 7.1 Vehicle Dispatching and Routing

This section reviews some basic concepts in vehicle dispatching and routing. Section 7.1.1 describes the instance data for these problems and section 7.1.2 specifies the concept of the *routing plan*, which is the output of traditional optimization algorithms for these problems. Section 7.1.3 specifies the arrival and departure times that can be derived from routing times. Finally, section 7.1.4 specifies the vehicle dispatching and routing problems considered in this book.

### 7.1.1 The Input Data

A vehicle dispatching or routing problem is specified by a number of customers that must be visited by a pool of vehicles. Each customer makes a request that must be served within a time window and takes some capacity from the vehicle. Each vehicle starts at its depot, serves some customers, and must return to its depot by the deadline.

**The Depot and the Customers** Each problem contains a set $R$ of $n$ customers and a set $O$ of $m$ depots. The set $S$ of sites is thus $R \cup O$. The travel time between sites $i$ and $j$ is denoted by $d(i,j)$.

These travel times satisfy the triangle inequality

$$\forall i, j, k \in S : d(i, k) \leq d(i, j) + d(j, k).$$

This inequality is natural in practice and typically enables the algorithms to be more efficient. It is not restrictive, however, and the results can be generalized when this assumption does not hold.

**The Requests**  Each request is associated with a customer and, since each customer makes at most one request per instance, we use the names *customer* and *request* interchangeably. Also, in the offline problems considered in this chapter, every customer makes exactly one request. Every request $c$ has a capacity $q(c) \geq 0$ and a service time $p(c) \geq 0$, which is the time to serve the request once the vehicle is on site.

**Vehicles**  Each instance has a pool of $m$ identical vehicles with capacity $Q$. Each vehicle starts from a different depot $o \in O$ and the algorithm may choose to deploy all of them or to use only a subset of them.[1]

**Time Windows**  Each customer $c$ has a time window specified by an interval $[e(c), l(c)]$ satisfying $e(c) \leq l(c)$. The time window represents the earliest and latest possible arrival times of a vehicle serving customer $c$. In other words, the service for customer $c$ may start as early as $e(c)$ and as late as $l(c)$. A customer $c$ may not be served before $e(c)$, but a vehicle arriving early to serve $c$ may wait at the site until time $e(c)$ before beginning service.

**Deadline**  All depots have a time window specified by an interval $[e_0, l_0]$, which represents the earliest departure and latest possible return for their vehicles. Typically, $e_0$ denotes the beginning of the day and $l_0$ is the deadline by which all vehicles must return to their depot. Observe also that depots have no service time.

### 7.1.2  Routing Plans

Optimization algorithms for vehicle dispatching and routing typically return a routing plan that specifies the order in which each vehicle visits its customers. The routing plan does not prescribe departure times for the vehicles but constrains them, as discussed in section 7.1.3.

---

[1]The fact that the vehicles start from different depots is purely a notational convenience. All the depots may have the same geographical location, in which case they are undistinguishable. Having a separate depot for each vehicle enables the algorithm to assign a single departure time to every site.

**Routes** A vehicle route, or *route* for short, starts at a depot, serves some customers, and returns to the depot. A customer appears at most once on a route. As a consequence, a route is a sequence

$$\langle o, c_1, \ldots, c_n, o \rangle$$

where $o \in O$, $c_i \in R$, and all $c_i$ are distinct. The capacity of a route $\rho$ is the sum of its customer capacities, that is,

$$q(\rho) = \sum_{i=1}^{n} q(c_i).$$

The travel time of a route $\rho$, denoted by $d(\rho)$, is the cost of visiting all its customers, that is,

$$d(\rho) = d(o, c_1) + d(c_1, c_2) + \ldots + d(c_{n-1}, c_n) + d(c_n, o).$$

**Routing Plan** A routing plan, or *plan* for short, is a tuple of routes

$$(\rho_1, \ldots, \rho_m),$$

one for each vehicle, in which each customer appears at most once. Because a customer makes exactly one request, a routing plan assigns a unique successor and predecessor for each served customer and depot. For a plan $\gamma$, the successor of site $c$ is denoted by $succ(c, \gamma)$ and the predecessor is denoted by $pred(c, \gamma)$. Since, in general, the discussion or definitions assume an underlying routing plan $\gamma$, we abuse notations and use $c^+$ and $c^-$ to denote the successor and predecessor of $c$ in $\gamma$. The travel time of a plan $\gamma = (\rho_1, \ldots, \rho_m)$, denoted by $d(\gamma)$, is the sum of the travel times of its routes, that is,

$$d(\gamma) = \sum_{i=1}^{m} d(\rho_i).$$

We also use $cust(\rho)$ and $cust(\gamma)$ to denote the customers of a route $\rho$ and a plan $\gamma$.

### 7.1.3 Service, Arrival, and Departure Times

Routing plans do not prescribe departure times for the vehicles. Moreover, the departure times of a routing plan typically are not uniquely defined: a vehicle may depart at different times from specific customers and still visit all its assigned customers before the deadline. However, the routing plan imposes constraints on the departure times. This section derives the earliest and latest departure times for each customer. Only earliest departure times are necessary to specify the offline problems. However, in an online context, the flexibility allowed by routing plans will be a tremendous asset. Most of the definitions use the traditional recursive equations [62].

**Earliest Service and Return Times** A route specifies an order for the service of its customers. From that order, it is possible to deduce the earliest start of service time for a customer with a routing plan. The *earliest departure time* of a site $i$, denoted by $\delta_i$, is defined recursively as

$$\begin{aligned}\delta(o) &= e_0 & (o \in O)\\ \delta(c) &= \max(\delta_{c^-} + d(c^-, c),\ e(c)) + p(c) & (c \in cust(\gamma)).\end{aligned}$$

Observe that this definition ensures that the vehicle serving customer $c$ waits until the beginning of $c$'s time window before starting service. The *earliest start of service time* of customer $c$, denoted by $\alpha(c)$, is defined as

$$\alpha(c) = \max(\delta(c^-) + d(c^-, c),\ e(c)) \quad (c \in cust(\gamma)).$$

The *earliest return time* of a route

$$\rho = \langle o, c_1, \ldots, v_n, o\rangle$$

is defined as

$$\alpha(\rho) = \begin{cases} \delta(c_n) + d(c_n, o) & \text{if } n > 0\\ e_0 & \text{otherwise.}\end{cases}$$

**Latest Arrival and Departure Times** The *latest arrival time* for a customer $c$, denoted by $z(c)$, is defined recursively as

$$\begin{aligned}z(o) &= l_0 & (o \in O)\\ z(c) &= \min(z(c^+) - d(c, c^+) - p(c),\ l(c)) & (c \in cust(\gamma)).\end{aligned}$$

The *latest departure time* for a site $s$, denoted by $\beta(s)$, is defined as

$$\beta(s) = z(s^+) - d(s, s^+).$$

### 7.1.4 Vehicle Routing and Dispatching Problems

We are now in position to describe the vehicle routing and dispatching considered in this book. A solution to a vehicle routing problem with time windows (VRPTW) is a routing plan $\gamma = (\rho_1, \ldots, \rho_m)$ satisfying the capacity and time window constraints, that is,

$$\begin{cases} q(\rho_j) \leq Q & (1 \leq j \leq m)\\ \alpha(\rho_j) \leq l_0 & (1 \leq j \leq m)\\ \alpha(c) \leq l(c) & (\forall c \in cust(\gamma)).\end{cases}$$

A solution to a vehicle dispatching problem is a routing plan $\gamma = (\rho_1, \ldots, \rho_m)$ that satisfies the capacity and deadline constraints, that is,

$$\begin{cases} q(\rho_j) \leq Q & (1 \leq j \leq m)\\ \alpha(\rho_j) \leq l_0 & (1 \leq j \leq m).\end{cases}$$

The objective is to find a solution maximizing the lexicographic function

$$w(\gamma) = (|cust(\gamma)|, -d(\gamma)).$$

This function maximizes the number of served customers $|cust(\gamma)|$ and, in case of ties, minimizes the total travel time. The difficulty in the dispatching problems of chapter 9 is the travel time, while maximizing the number of served customers is the main challenge in the vehicle routing problems of chapter 10. Observe also that this objective function differs from the optimization criterion used, for instance, in the Solomon benchmarks. In the Solomon problems, the goal is to minimize the number of vehicles and, in case of ties, to minimize the total travel time, which corresponds more to strategic planning than the operational decision making of online optimization.

## 7.2  Large Neighborhood Search

This section presents the optimization algorithm used in online vehicle dispatching and routing. The algorithm is a large neighborhood search (LNS) that produces high-quality solutions when the objective consists of maximizing the number of served customers and minimizing travel time. It is taken from an effective two-stage algorithm [9] that uses a simulated annealing approach to minimize the number of vehicles and LNS to minimize the total travel time. LNS can be viewed as a variable-neighborhood local search [49] in which a move consists of relocating a number of customers. The relocation is performed by an incomplete search procedure based on discrepancy search. Section 7.2.1 specifies the neighborhood, section 7.2.2 presents the LNS algorithm, and section 7.2.3 discusses how to relocate customers.

### 7.2.1  The Neighborhood

Given a solution $\gamma$, the neighborhood of the LNS algorithm, denoted by $\mathcal{N}_R(\gamma)$, is the set of solutions that can be reached from $\gamma$ by relocating at most $k$ customers (where $k$ is a parameter of the implementation). Since the LNS algorithm also uses subneighborhoods and explores the neighborhood in a specific order, it is convenient to introduce some additional notations. In particular, $\mathcal{N}_R(\gamma, S)$ denotes the set of solutions that can be reached from $\gamma$ by relocating the customers in $S$. Moreover, if $\gamma$ is a partial solution not serving any customer in $S$, then $\mathcal{N}_I(\gamma, S)$ denotes the solutions obtained by inserting the customers $S$ in $\gamma$.

### 7.2.2  The LNS Algorithm

At a high level, LNS can be seen as a local search where each iteration selects a neighbor

$$\gamma_c \in \mathcal{N}_R(\gamma_b)$$

LNS($\gamma_b$)
```
 1  for l ← 1...maxSearches do
 2     for n ← 1...p do
 3        i ← 1;
 4        while i ≤ maxIterations do
 5           i ← i + 1;
 6           S ← SELECTCUSTOMERS(cust(γb), n);
 7           γc ← arg-min(γ ∈ NR(γb, S)) w(γ);
 8           if w(γc) < w(γb) then
 9              γb ← γc;
10              i ← 1;
```

**Figure 7.1:** The LNS Algorithm for Vehicle Routing and Dispatching.

and accepts the move if $w(\gamma_c) < w(\gamma_b)$. It can be formalized as follows:

LNS()
```
 1  for i ← 1...maxIterations do
 2     SELECT γc ∈ NR(γb);
 3     if w(γc) < w(γb) then
 4        γb ← γc;
```

In practice, it is beneficial to incorporate three additional ideas into the algorithm.

1. The algorithm should start by relocating small sets of customers and progressively increase the number of relocations when no improvement has taken place, instead of systematically relocating $k$ customers.

2. The algorithm should be more greedy and select the best relocation for a randomly chosen subset of customers.

3. The local search is iterated multiple times.

The LNS algorithm is depicted in figure 7.1. The algorithm performs a number of local search (line 1). Each local search successively considers relocations with 1, 2, ..., $k$ customers (line 2). For a fixed number $n$ of relocations, the LNS algorithm attempts *maxIterations* relocations. Each of these attempts randomly selects a set $S$ of customers from $cust(\gamma_b)$ (line 6) and determines the best way to relocate them (line 7). If the resulting plan is better than the best existing plan (line 8), the relocation is accepted (line 9) and the number of iterations with $n$ relocations is reset (line 10).

SELECTCUSTOMERS($Customers, n$)
1   $S \leftarrow \{\text{RANDOM}(Customers)\};$
2   **for** $i \leftarrow 2 \ldots n$ **do**
3       $c \leftarrow \text{RANDOM}(S);$
4       LET $\{c_0, \ldots, c_{N-i}\} = Customers \setminus S$
5           WHERE $\forall 1 \leq i < j \leq N : \text{RELATEDNESS}(c, c_i) \leq \text{RELATEDNESS}(c, c_j);$
6       $r \leftarrow \lfloor \text{RANDOM}([0, 1])^\theta \times |Customers \setminus S| \rfloor;$
7       $S \leftarrow S \cup \{c_r\};$

**Figure 7.2:** Selecting Customers in the LNS Algorithm.

### 7.2.3 Relocation

It remains to describe how to select customers and how to implement line 7 in the LNS algorithm.

**Selecting Customers to Relocate**   The implementation of SELECTCUSTOMERS is depicted in figure 7.2. Its key idea is to select customers that are relatively close geographically. More precisely, at each iteration, the algorithm maintains a set $S$ of already selected customers. To add a new customer to relocate, function SELECTCUSTOMERS selects a random customer $c$ in $S$ (line 3) and ranks all remaining customers with respect to their proximity to $c$ (lines 4 and 5). The new customer $c_r$ is then selected randomly with a probability that favors related customers (lines 6 and 7). The relatedness measure in line 5 is defined as

$$\text{RELATEDNESS}(i, j) = \frac{1}{d'(i, j) + v(i, j)}$$

where

$$d'(i, j) = d(i, j) \; / \max_{i,j \in S} d(i, j)$$

and

$$v(i, j) = \begin{cases} 0 & \text{if customers } i \text{ and } j \text{ are on the same route} \\ 1 & \text{otherwise.} \end{cases}$$

In other words, the relatedness measure is inversely proportional to the normalized travel times with a bias for customers on the same vehicle. Note also that $\theta$ is called the deterministic factor: the greater it is, the more deterministic the algorithm.

**The Exploration Algorithm**   The LNS algorithm uses a branch and bound algorithm to explore the selected subneighborhood. The algorithm is depicted in figure 7.3. It receives as inputs the current plan $\gamma_c$, the set $S$ of customers to insert, and the best solution found so far $\gamma^*$. If the set of

DFSEXPLORE$(\gamma_c, S, \gamma^*)$
1   **if** $S = \emptyset$ **then**
2     **if** $w(\gamma_c) < w(\gamma^*)$ **then**
3       $\gamma^* \leftarrow \gamma_c$;
4   **else**
5     $c \leftarrow \text{arg-max}(c \in S) \;\; \text{arg-min}(\gamma \in \mathcal{N}_I(\gamma, \{c\})) \;\; w(\gamma)$;
6     $S_c \leftarrow S \setminus \{c\}$;
7     $\langle \gamma_1, \ldots, \gamma_k \rangle = \mathcal{N}_I(\gamma, \{c\})$   WHERE   $\forall 1 \leq i \leq j \leq k : w(\gamma_i) \leq w(\gamma_j)$;
8     **for** $i \leftarrow 1 \ldots k$ **do**
9       **if** BOUND$(\gamma_i, S_c) < w(\gamma^*)$ **then**
10        DFSEXPLORE$(\gamma_i, S_c, \gamma^*)$;

**Figure 7.3:** The Branch and Bound Algorithm for the Neighborhood Exploration in LNS.

customers to insert is empty, the algorithm checks whether the current solution improves the best solution found so far (lines 1 through 3). Otherwise, it selects the customer whose best insertion degrades the objective function the most (line 5). The algorithm then explores all insertion points for customer $c$ in increasing order of their travel times (lines 8 through 10). Observe that only the partial plans whose lower bounds are better than the best solution $\gamma^*$ are explored by the algorithm. The lower bound satisfies the inequality

$$\text{BOUND}(\gamma, S) \leq \min_{\gamma' \in \mathcal{N}_I(\gamma, S)} w(\gamma').$$

It remains to discuss the lower bound and how to keep the computation times reasonable.

**Bounding**   To evaluate the travel time component of the objective function, the bounding function computes the cost of a minimum spanning k-tree [37] on the insertion graph with the depot as the distinguished vertex, thus generalizing the well-known 1-tree bound of the traveling salesman problem. The insertion graph for a plan $\gamma$ is defined as follows: The vertices are the customers and the edges come from three different sets:

1. The edges coming from the routes in $\gamma$;

2. all the edges between customers in $S$;

3. all the feasible edges connecting a customer from $cust(\gamma)$ and a customer from $S$.

More precisely, the insertion graph is defined as follows.

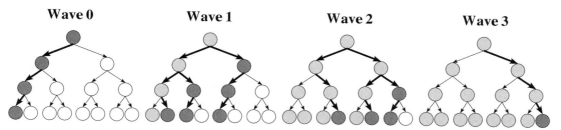

**Figure 7.4:** The Successive Waves of LDS.

**Definition 7.1 (Insertion Graph)** Let $\gamma$ be a partial plan over customers $R$ and $S$ be the set of customers to insert. The insertion graph is the graph $G(R \cup S, E)$ where

$$
\begin{aligned}
E &= E_\gamma \cup E_S \cup E_c; \\
E_\gamma &= \{(i, i^+) \mid i \in R\}; \\
E_S &= \{(i, j) \mid i, j \in S\}; \\
E_c &= \{(pred(j, \gamma'), j) \mid j \in S \,\&\, \gamma' \in \mathcal{N}_I(\gamma, \{j\}) \,\&\, pred(j, \gamma') \in R\} \;\cup \\
&\quad\; \{(j, succ(j, \gamma')) \mid j \in S \,\&\, \gamma' \in \mathcal{N}_I(\gamma, \{j\}) \,\&\, succ(j, \gamma') \in R\}.
\end{aligned}
$$

**Incomplete Search**  For large numbers of customers, finding the best reinsertion may be too time-consuming and the LNS borrows a strategy from limited discrepancy search (LDS) to explore only a small part of the search tree.

Limited discrepancy search [50] is a search strategy that assumes the problem at hand has a good heuristic. Its basic idea is to explore the search tree in waves, and each successive wave allows the heuristic to make more mistakes. Wave 0 simply follows the heuristic. Wave 1 explores the solutions that can be reached by assuming that the heuristic made one mistake. More generally, wave $i$ explores the solutions that can be reached by assuming that the heuristic makes $i$ mistakes. Figure 7.4 illustrates these waves graphically on a binary tree. The figure describes the successive waves used in exploring the tree. The nodes visited in a given wave are colored black and those visited in previous waves are colored grey. By exploring the search tree according to the heuristic, LDS may reach good solutions (and thus an optimal solution) much faster than depth-first and best-first search for some applications. Its strength is its ability to explore diverse parts of the search tree containing good solutions that are reached only much later by depth-first search.

The LNS implementation uses one phase of limited discrepancy search that allows up to $d$ discrepancies. Figure 7.5 depicts the algorithm. Observe that, in the LNS algorithm, the tree is not binary and the heuristic selects the insertion points by increasing lower bounds.

$\text{LDSEXPLORE}(\gamma_c, S, \gamma^*, d)$

```
 1   if d < 0 then
 2      return ;
 3   if S = ∅ then
 4      if w(γc) < w(γ*) then
 5         γ* ← γc;
 6   else
 7      c ← arg-max(c ∈ S)  arg-min(γ ∈ NI(γ, {c}))  w(γ);
 8      Sc ← S \ {c};
 9      ⟨γ1, . . . , γk⟩ = NI(γ, {c})   WHERE   ∀1 ≤ i ≤ j ≤ k : w(γi) ≤ w(γj);
10      for i ← 1 . . . k do
11         if BOUND(γi, Sc) < w(γ*) then
12            LDSEXPLORE(γi, Sc, γ*, d);
13            d ← d − 1;
```

**Figure 7.5:** The Branch and Bound Algorithm with a Limited Discrepancy Strategy.

## 7.3   Notes and Further Reading

Large neighborhood search was initially proposed in [102], which also proposed the relatedness measure and the branching heuristic discussed here. LNS was later used in [9] to minimize travel times in a two-stage hybrid local search algorithm. The first stage uses simulated annealing to reduce the number of vehicles, and the resulting algorithm improved or matched most of the best-known solutions to the Solomon benchmarks when it was proposed. Very few of the Solomon benchmarks [105] (which involves 100 customers) have been solved to optimality. Readers can consult [37, 65] for recent results. One of the advantages of LNS is its flexibility in accommodating side constraints that often arise in industrial applications. Reference [15] demonstrates that LNS could be successfully used for pickup and delivery problems with time windows.

# 8 Online Stochastic Routing

*I have yet to see any problem, however complicated, which, when you looked at it
in the right way, did not become still more complicated*
— Poul Anderson

This chapter presents the online stochastic algorithms for multiple vehicle routing problems. These
applications are significantly more challenging than the packet scheduling and multiknapsack problems presented earlier. Besides the inherent complexity of the underlying offline problems, they
involve a number of complex issues such as service guarantees, multiple joint decisions, and the
fact that serving a request also induces travel times. Some of these difficulties are addressed by
innovative techniques such as waiting and relocation strategies. This chapter first specifies the online problems and identifies some fundamental issues. It then presents the algorithms in stepwise
refinements starting with single vehicle routing before moving to waiting and relocation strategies,
and multiple vehicle routing.

## 8.1  Online Stochastic Vehicle Routing

In online stochastic problems, customers arrive dynamically as the algorithm proceeds. Each customer request has a time window (possibly the entire time horizon) during which it can be served
and, obviously, a request cannot be served before it occurs. Upon arrival, the algorithm must decide
whether to accept or reject the request. If the request is accepted, the online algorithm must serve
it before the time horizon. However, the online algorithm must not commit a vehicle nor commit to
a specific order within the routing plan upon acceptance. As a consequence, online vehicle routing
provides another entry point in the spectrum of online algorithms, and it is useful to contrast it
with online packet scheduling and multiknapsack with deadlines.

1. In packet scheduling, the online algorithm does not decide whether to accept or reject packets.
   Packets are simply dropped once their deadlines expire.

2. In online multiknapsack with deadlines, the algorithm must decide whether to accept or reject
   a request. Moreover, upon acceptance, the online algorithm must determine an allocation for
   the request and this allocation is final.

3. In online routing, the algorithm must also decide whether to accept or reject a request, but it
   does not need to select a vehicle, a starting time, or a predecessor for the customer. It must
   simply guarantee that the customer will be served in the routing plan.

Online stochastic routing problems are much more challenging than earlier applications in this book.
Besides the inherent complexity of the underlying offline optimization problems, they raise a number
of fundamental difficulties, some of which are reviewed here.

1. Online routing problems typically have a large number of choices. A typical application may consist of eighty requests available at the beginning of the day and eighty additional requests arriving dynamically over the course of the algorithm. A decision at a given step must sometimes be made among a large number of requests (for instance, eighty), while the time within decisions allows for only 3 to 5 optimizations. As a consequence, algorithm $\mathcal{E}$ is completely impractical because the time to evaluate each decision on a single scenario would require already eighty optimizations.

2. Contrary to earlier applications, serving a request takes more than a single unit of time in online routing. Indeed the vehicle must travel to the customer from its current location and serve the request. As a result, no other request can be served until the vehicle becomes idle and is ready to serve another customer.

3. Since serving a customer requires multiple units of time, solutions to the scenario may schedule "sampled" customers (in contrast to actual requests) first on a vehicle. It is a fundamental issue to decide how best to proceed in such circumstances. This chapter presents waiting and relocation strategies to take advantage of such opportunities and to improve the quality of the routings significantly.

4. The online routing algorithm must not only determine a high-quality routing plan; contrary to the offline problem, it must also determine the departure time online, since the customers must be visited as the algorithm proceeds. Once again, there are design decisions related to the departure times that may affect the quality of the routing plan significantly.

5. Online routing applications face the additional difficulty of handling multiple vehicles. First, several vehicles may be idle at some point and must be allocated to a new request. Second, even if only one vehicle is idle, future decisions on the other vehicles may have a fundamental impact on the current selection for the idle vehicle.

**Notations** This chapter uses some additional notations on sequences. If $S$ is a nonempty sequence, FIRST$(S)$ and LAST$(S)$ denote the first and the last element of a nonempty sequence. If $S$ is a sequence and $S^-$ is a prefix of $S$, then $S - S^-$ denotes the suffix $S^+$ such that $S = S^- : S^+$. If $S$ is a sequence and $R$ is a set, FILTER$(S, R)$ denotes the sequence $S$ where the elements of $R$ have been removed from $R$, that is,

$$
\begin{aligned}
\text{FILTER}(\langle\rangle, R) \quad &= \quad \langle\rangle \\
\text{FILTER}(\langle r\rangle : S, R) \quad &= \quad \begin{cases} \langle r\rangle : \text{FILTER}(S, R) & \text{if } r \notin R \\ \text{FILTER}(S, R) & \text{otherwise.} \end{cases}
\end{aligned}
$$

The offline routing problems can be specified abstractly to accommodate problem-specific constraints beyond the time windows and the capacity constraints presented earlier. The offline problem can be seen then as the finding of a routing plan $\gamma$ maximizing the lexicographic objective function

$$w(\gamma) = (|cust(\gamma)|, -d(\gamma))$$

while satisfying the problem-specific constraints $C(\gamma)$. The online algorithm maintains a partial routing plan

$$\gamma^- = \langle \rho_1^-, \ldots, \rho_m^- \rangle$$

consisting a partial route $\rho_i$ for each vehicle $i$. It also assigns departure times to all customers in

$$cust(\gamma^-) \setminus \{\text{LAST}(\rho_1), \ldots, \text{LAST}(\rho_m)\}.$$

These departure times specify when the vehicles serving a given customer departed to the next customer. The last customers on the vehicles have no departure times, since they have not been served and their successors are not known.

The online algorithms have at their disposal an optimization algorithm $\mathcal{O}$ that, given a pair $(\gamma^-, \sigma^-)$ where $\gamma^-$ represents the partial routing plan and $\sigma^-$ contains its associated departure times, returns a routing plan $\gamma^*$ that extends $\gamma^-$, minimizes the objective function, and satisfies the problem-specific constraints. More precisely, assume that the departure times are specified by a partial assignment $\sigma : Customers \to H$ and that $C[\sigma^-]$ denotes the problem-specific constraints $C$ where the time windows of each customer $c$ in

$$cust(\gamma^-) \setminus \{\text{LAST}(\rho_1), \ldots, \text{LAST}(\rho_m)\}$$

have been tightened to

$$[\sigma^-(c), \sigma^-(c)].$$

Given a set of customer requests $R$ and a pair $\gamma^-, \sigma^-$ where

$$\gamma^- = \langle \rho_1^-, \ldots, \rho_m^- \rangle,$$

the optimization algorithm $\mathcal{O}(\gamma^-, \sigma^-, R)$ returns a routing plan $\gamma^+$ where

$$\gamma^+ = \langle \rho_1^- : \rho_1^+, \ldots, \rho_m^- : \rho_m^+ \rangle$$

minimizing $w$ and satisfying $C[\sigma^-]$.

The rest of this chapter describes the online stochastic algorithms in stepwise refinements, focusing on a single issue at a time. Section 8.2 describes the online stochastic algorithms for a single vehicle. Section 8.3 shows how to incorporate service guarantees in the algorithms. Sections 8.4 and 8.5 introduce waiting and relocation strategies that may significantly improve the quality of online routing plans. Section 8.6 presents the online stochastic algorithms for multiple vehicles and introduces the idea of pointwise decisions.

ONLINE ALGORITHM $\mathcal{A}(\langle R_1, \ldots, R_h \rangle)$
1   $\rho_0 \leftarrow \langle \rangle$;
2   $\sigma_0 \leftarrow \sigma_\perp$;
3   $\Gamma \leftarrow \text{GENERATESOLUTIONS}(\rho_0, \sigma_0, R_1)$;
4   **for** $t \in H$ **do**
5     **if** $\text{IDLE}(\rho_{t-1}, \sigma_{t-1})$ **then**
6       $s_t \leftarrow \text{CHOOSEREQUEST}(\rho_{t-1}, \boldsymbol{R_t}, \Gamma)$;
7       $\rho_t \leftarrow \rho_{t-1} : s_t$;
8       $\sigma_t \leftarrow \sigma_{t-1}[\text{LAST}(\rho_{t-1}) \leftarrow t]$;
9       $\Gamma \leftarrow \{ \rho \in \Gamma \mid \text{FIRST}(\rho - \rho_{t-1}) = s_t \}$;
10    **else**
11      $\rho_t \leftarrow \rho_{t-1}$;
12      $\sigma_t \leftarrow \sigma_{t-1}$;
13    $\Gamma \leftarrow \Gamma \cup \text{GENERATESOLUTIONS}(\rho_t, \sigma_t, \boldsymbol{R_t})$;
14  **return** $(\rho_h, \sigma_h)$;

GENERATESOLUTIONS$(\rho_t, \sigma_t, \boldsymbol{R_t})$
1   $\Gamma \leftarrow \emptyset$;
2   **repeat**
3     $A \leftarrow \boldsymbol{R_t} : \text{SAMPLE}(t, h)$;
4     $\Gamma \leftarrow \Gamma \cup \{\mathcal{O}(\rho_t, \sigma_t, A)\}$;
5   **until** time $t + 1$
6   **return** $\Gamma$;

**Figure 8.1:** Online Stochastic Routing with Precomputation

## 8.2   Online Single Vehicle Routing

We now present the generic online routing algorithm for a single vehicle. Since there is only one vehicle, a routing plan is simply a vehicle route and we use both terms interchangeably in the next two sections. The algorithm is depicted in figure 8.1 and includes precomputation since, in practical applications, it is between decision times that optimizations are available. It maintains a set of plans $\Gamma$ to make decisions over the course of the computation. At every time $t$, the algorithm also maintains a partial routing plan $\rho_t$ and its associated departure times $\sigma_t$. The algorithm also assumes that the set of requests $R_1$ is available before the start of the computation, as is typical in practical applications (for instance, $R_1$ may be the customers who called late the day before).

Lines 1 and 2 initialize the partial routing plan and the departure times, while line 3 generates the initial set of plans used in the decisions. The body of the algorithm (lines 5 through 13) first

CHOOSEREQUEST-$\mathcal{C}(\rho_t, \boldsymbol{R_t}, \Gamma)$
1    $F \leftarrow \bigcup_{i=1}^{t} R_i$;
2    **for** $r \in F$ **do**
3       $f(r) \leftarrow 0$;
4    **for** $\rho \in \Gamma$ **do**
5       $r \leftarrow$ FIRST(FILTER$(\rho - \rho_t, F)$);
6       $f(r) \leftarrow f(r) + 1$;
7    **return** $argmax(r \in F)\ f(r)$;

**Figure 8.2:** The Consensus Algorithm for Online Stochastic Routing.

determines whether the vehicle is idle, that is, whether service is completed for the last customer in $\rho_{t-1}$ given the departure times in $\sigma_{t-1}$. If the vehicle is busy traveling or servicing the last customer in $\rho_{t-1}$, the routing plan and departure times remain the same (lines 11 and 12) and the algorithm simply continues generating plans (line 13). Otherwise, the vehicle is idle and the algorithm chooses a request $s_t$ to serve using the plans in $\Gamma$ (line 6), augments the routing plan (line 7) and the departure times (line 8), and updates $\Gamma$ to remove the plans incompatible with the decisions (line 9). It is useful to review some of the detail of the algorithm.

- A vehicle is idle at time $t$ for a routing plan $\langle s_1, \ldots, s_k \rangle$ and departure times $\sigma$ if

$$\text{IDLE}(\langle s_1, \ldots, s_k \rangle, \sigma) \equiv k = 0 \ \vee \ \max(\sigma(s_{k-1}) + d(s_{k-1}, s_k), e(s_k)) + p(s_k) \leq t.$$

  In other words, a vehicle is idle if its route has no customer or if it has served its last customer $s_k$ at time $t$, given that it departed from customer $s_{k-1}$ at time $\sigma(s_{k-1})$, traveled for $d(s_{k-1}, s_k)$ units of time, and served $s_k$ for $p(s_k)$ time units starting no earlier than $e(s_k)$.

- The algorithm assigns the departure time of the last customer in $\rho_{t-1}$ to time $t$ in line 8. This means that the algorithm departs for customer $s_t$ at time $t$.

- A routing plan $\rho$ that visits a customer other than $s_t$, that is,

$$\text{FIRST}(\rho - \rho_{t-1}) \neq s_t$$

  must be removed from $\Gamma$ since its decisions are incompatible with $\rho_t$ and cannot be used for future decision steps.

Figure 8.1 also depicts how to generate plans for immediate decision making. Line 3 of function GENERATESOLUTIONS generates a scenario by sampling the distribution from time $t$ to the horizon. Line 4 calls the optimization algorithm with the routing plan and departure times at time $t$.

Figure 8.2 shows how to implement function SELECTREQUEST to obtain the consensus algorithm $\mathcal{C}$. Algorithm $\mathcal{C}$ considers all known requests $F$ (line 1) and initializes their evaluations (lines 2 and

3). It then considers each routing plan $\rho \in \Gamma$ (line 4), retrieves the request served next in $\rho$, and increments its credit (line 6). The request in $F$ with the best evaluation is selected in line 7.

We must emphasize a critical point in this implementation. A solution $\rho \in \Gamma$ is a routing plan

$$\rho = \rho_{t-1} : \rho^+$$

starting with partial route $\rho_{t-1}$ followed by a sequence of requests coming from $F$ and the sampling. As a consequence, and contrary to the applications seen earlier, there is no guarantee that the request $s$ served next on the vehicle, that is,

$$s = \text{FIRST}(\rho^+),$$

is an actual request ($s \in F$), not a sampled customer (w$s \notin F$). This is precisely why the implementation in figure 8.2 uses

$$\text{FILTER}(\rho^+, F)$$

to prune plan $\rho^+$ and keep only the requests in $F$. This guarantees that

$$\text{FIRST}(\text{FILTER}(\rho^+, F))$$

returns a real customer and that the vehicle departs for a customer who actually requested service.

## 8.3  Service Guarantees

Typical online vehicle routing applications feature service guarantees. The algorithm may decide to accept or reject a new request, but, whenever a request is accepted, the request must be served before the time horizon. Note that, upon acceptance, the algorithm promises only that the request will be served; it must not specify when it will be served. Figure 8.3 depicts the generic online algorithm with service guarantees. The main modifications are in lines 5 and 6. Function ACCEPTREQUESTS decides which request to accept, while function UPDATEPLANS updates the plan to accommodate the newly accepted requests. If a plan cannot include the new requests, it is removed from $\Gamma$. It is important to stress that at least a plan $\rho \in \Gamma$ should be able to accommodate the requests $A_t$ since otherwise the algorithm cannot provide the necessary service guarantees. The rest of the algorithm is essentially left unchanged, but uses the accepted requests $\boldsymbol{A_t}$ instead of the actual requests $\boldsymbol{R_t}$.

In the presence of service guarantees, it is necessary to generalize the optimization algorithm that now receives the set $A$ of accepted requests and the set $R$ of sampled customers. A feasible solution given a partial routing plan $\gamma_t$ and departure times $\sigma_t$ returns a solution $\gamma = \gamma_t : \gamma^+$ such that

$$C[\sigma_t](\gamma) \ \wedge \ A \subseteq cust(\gamma)$$

and the optimization function $w(\gamma)$ is maximized. Observe that this is precisely the optimization problem discussed in chapter 7.

ONLINE ALGORITHM $\mathcal{A}(\langle R_1, \ldots, R_h \rangle)$
1   $\rho_0 \leftarrow \langle \rangle$;
2   $\sigma_0 \leftarrow \sigma_\perp$;
3   $\Gamma \leftarrow \text{GENERATESOLUTIONS}(\rho_0, \sigma_0, R_1)$;
4   **for** $t \in H$ **do**
5     $A_t \leftarrow \text{ACCEPTREQUESTS}(\rho_{t-1}, \sigma_{t-1}, R_t, \boldsymbol{A_{t-1}}, \Gamma)$;
6     $\Gamma \leftarrow \text{UPDATEPLANS}(\rho_{t-1}, \sigma_{t-1}, \boldsymbol{A_t}, \Gamma)$;
7     **if** $\text{IDLE}(\rho_{t-1}, \sigma_{t-1})$ **then**
8       $s_t \leftarrow \text{CHOOSEREQUEST}(\rho_{t-1}, \boldsymbol{A_t}, \Gamma)$;
9       $\rho_t \leftarrow \rho_{t-1} : s_t$;
10     $\sigma_t \leftarrow \sigma_{t-1}[\text{LAST}(\rho_{t-1}) \leftarrow t]$;
11     $\Gamma \leftarrow \{ \rho \in \Gamma \mid \text{FIRST}(\rho - \rho_{t-1}) = s_t \}$;
12     **else**
13       $\rho_t \leftarrow \rho_{t-1}$;
14       $\sigma_t \leftarrow \sigma_{t-1}$;
15     $\Gamma \leftarrow \Gamma \cup \text{GENERATESOLUTIONS}(\rho_t, \sigma_t, \boldsymbol{A_t})$;
16  **return** $(\rho_h, \sigma_h)$;

GENERATESOLUTIONS$(\rho_t, \sigma_t, \boldsymbol{A_t})$
1  $\Gamma \leftarrow \emptyset$;
2  **repeat**
3    $R \leftarrow \text{SAMPLE}(t, h)$;
4    $\Gamma \leftarrow \Gamma \cup \{\mathcal{O}(\rho_t, \sigma_t, \boldsymbol{A_t}, R)\}$;
5  **until** time $t + 1$
6  **return** $\Gamma$;

**Figure 8.3:** Online Stochastic Routing with Precomputation and Service Guarantees

It is also useful to discuss possible implementations of function ACCEPTREQUESTS. To simplify the discussion, consider the case where only one request $r$ arrives at time $t$. A greedy implementation of ACCEPTREQUESTS would determine whether a plan $\rho \in \Gamma$ can accommodate request $r$, possibly by removing some sampled customer. The request is accepted if such a plan exists and rejected otherwise. More sophisticated implementations may evaluate the average number of sampled customers that must be removed to accommodate the requests and accept the request when this number is no greater than one. These strategies are evaluated experimentally in chapter 10.

$\textsc{chooseRequest-}\mathcal{C}(\rho_t, \boldsymbol{A_t}, \Gamma)$

```
1   F ← ∪ᵗᵢ₌₁ Aᵢ;
2   for r ∈ F ∪ {⊥} do
3       f(r) ← 0;
4   for ρ ∈ Γ do
5       r ← FIRST(ρ − ρₜ);
6       if r ∈ F then
7           f(r) ← f(r) + 1;
8       else
9           f(⊥) ← f(⊥) + 1;
10  return argmax(r ∈ F ∪ {⊥}) f(r);
```

**Figure 8.4:** The Consensus Algorithm for Online Stochastic Routing with a Waiting Strategy.

## 8.4   A Waiting Strategy

One of the challenges in online vehicle routing is the possibility that the customer served next in a scenario may be a sampled customer. The implementation of algorithm $\mathcal{C}$ in figure 8.2 solves this difficulty by filtering sampled customers before selecting the request. This section investigates a waiting strategy to address the same problem. It is based on the recognition that, in some circumstances, it may be beneficial for the vehicle to wait at its current location instead of serving customers too eagerly. For instance, the fact that the solution $\rho$ to a scenario at time $t$ starts with a sampled customers, that is,

$$\rho = \rho_{t-1} : \rho^+ \ \wedge \ \textsc{First}(\rho^+) \notin F,$$

indicates that it may be beneficial to wait since the sampled request may materialize, in which case it must be served before the first accepted customer. The difficulty is deciding when to wait in a systematic fashion given that the algorithm has solved multiple scenarios, all of which may have different customers to serve next in their routing plans. Figure 8.4 depicts a natural implementation. Its key idea is to add a wait action $\perp$ to the accepted requests. When considering a plan $\rho \in Gamma$, the algorithm retrieves the request $r$ to serve next in the scenario (line 5). In the case of an accepted request ($r \in F$), the evaluation of $r$ is incremented. Otherwise, if $r$ is a sampled customer ($r \notin F$), the evaluation of the wait action is incremented. The implementation then selects the elements of $F \cup \{\perp\}$ with the best evaluation, which may be either an accepted request or the wait action.

The online generic routing algorithm must also be generalized to wait. When the request is the wait action, the algorithm modifies neither the routing plan nor the departure times.

ONLINE ALGORITHM $\mathcal{A}(\langle R_1, \ldots, R_h \rangle)$
1   $\rho_0 \leftarrow \langle\rangle$;
2   $\sigma_0 \leftarrow \sigma_\perp$;
3   $\Gamma \leftarrow \text{GENERATESOLUTIONS}(\rho_0, \sigma_0, R_1)$;
4   **for** $t \in H : t > 1$ **do**
5      $A_t \leftarrow \text{ACCEPTREQUESTS}(\rho_{t-1}, \sigma_{t-1}, R_t, \boldsymbol{A_{t-1}}, \Gamma)$;
6      $\Gamma \leftarrow \text{UPDATEPLANS}(\rho_{t-1}, \sigma_{t-1}, \boldsymbol{A_t}, \Gamma)$;
7      $s_t \leftarrow \perp$;
8      **if** $\text{IDLE}(\rho_{t-1}, \sigma_{t-1})$ **then**
9         $s_t \leftarrow \text{CHOOSEREQUEST}(\rho_{t-1}, \boldsymbol{A_t}, \Gamma)$;
10    **if** $s_t \neq \perp$ **then**
11      $\rho_t \leftarrow \rho_{t-1} : s_t$;
12      $\sigma_t \leftarrow \sigma_{t-1}[\text{LAST}(\rho_{t-1}) \leftarrow t]$;
13      $\Gamma \leftarrow \{ \rho \in \Gamma \mid \text{FIRST}(\rho - \rho_{t-1}) = s_t \}$;
14    **else**
15      $\rho_t \leftarrow \rho_{t-1}$;
16      $\sigma_t \leftarrow \sigma_{t-1}$;
17    $\Gamma \leftarrow \Gamma \cup \text{GENERATESOLUTIONS}(\rho_t, \sigma_t, \boldsymbol{A_t})$;
18   **return** $(\rho_h, \sigma_h)$;

**Figure 8.5:** The Online Stochastic Routing with Precomputation, Service Guarantees, and a Waiting Strategy.

## 8.5  A Relocation Strategy

The waiting strategy recognizes that it may be beneficial to wait at the current location instead of serving an accepted request. It is especially appropriate for problems in which the bottleneck is to minimize travel times and it is reasonably easy to serve the customers. When the challenge is in maximizing the number of served requests, it is appealing to explore a relocation strategy and to consider moving to the location of sampled customers. Once again, the difficulty is determining when and where to move. Figure 8.6 proposes a natural relocation strategy. Its fundamental idea is to avoid differentiating between accepted and sampled customers: the vehicle simply moves to the request with the best evaluation. In the implementation, lines 1 and 2 initialize the evaluation of all customers, lines 4 and 5 increments the first request, and line 6 selects the request with the best evaluation. The selected request may be either an accepted or a sampled request.

Note that a relocation strategy may be beneficial for improving the number of served requests because it anticipates future requests and positions the vehicle to serve them quickly. It is never advantageous when minimizing travel times, since it may move to locations where no requests will ever materialize.

CHOOSEREQUEST-$\mathcal{C}(\rho_t, \boldsymbol{A_t}, \Gamma)$
1  **for** $r \in Customers$ **do**
2      $f(r) \leftarrow 0$;
3  **for** $\rho \in \Gamma$ **do**
4      $r \leftarrow \text{FIRST}(\rho - \rho_t)$;
5      $f(r) \leftarrow f(r) + 1$;
6  **return** $argmax(r \in Customers) \, f(r)$;

**Figure 8.6:** The Consensus Algorithm for Online Stochastic Routing with a Relocation Strategy.

## 8.6 Multiple Pointwise Decisions

It remains to show how to generalize the generic online routing algorithm to multiple vehicles. A solution is a routing plan $\gamma = (\rho_1, \ldots, \rho_m)$, where $\rho_i$ is the route of vehicle $i$. The introduction of multiple vehicles raises a number of issues.

1. There is a routing decision each time one of the vehicles is idle. So the definition of function IDLE becomes

$$\text{IDLE}(\gamma, \sigma) \equiv \bigvee_{i=1}^{m} \text{IDLE}(\gamma(i), \sigma).$$

2. There may be several simultaneous decisions whenever several vehicles are idle at the same time.

3. Even if only one vehicle is idle, the routing on other vehicles have an impact on the decision for the idle vehicle and, as a result, the algorithm should consider all vehicles together when making decisions.

4. The evaluation of the decisions is more challenging since each decision corresponds to a tuple $(r_1, \ldots, r_m)$ representing the customer to schedule next on each vehicle.

The generic online stochastic routing algorithm for multiple vehicles is depicted in figure 8.7. The main modifications are located in lines 8 through 16. Function CHOOSEREQUEST in line 8 returns a tuple of requests

$$ts = (s_1, \ldots, s_m),$$

where $s_i$ represents the customer to schedule next on vehicle $i$. Observe that the vehicle may not be able to depart for this customer at time $t$, since it may be busy traveling or serving the last customer in $\gamma_{t-1}(i)$. Lines 10 through 15 consider each idle vehicle $i$ for which the decision $ts(i)$ is not the waiting action. Scheduling a request $ts(i)$ consists of assigning a departure time (line 13), concatenating $ts(i)$ to route $\rho_i$ (line 14), and pruning the plans incompatible with this decision (line

ONLINE ALGORITHM $\mathcal{A}(\langle R_1, \ldots, R_h \rangle)$
1   $\gamma_0 \leftarrow \langle \rangle$;
2   $\sigma_0 \leftarrow \sigma_\perp$;
3   $\Gamma \leftarrow$ GENERATESOLUTIONS$(\gamma_0, \sigma_0, R_1)$;
4   **for** $t \in H$ **do**
5       $A_t \leftarrow$ ACCEPTREQUESTS$(\rho_{t-1}, \sigma_{t-1}, R_t, \boldsymbol{A_{t-1}}, \Gamma)$;
6       $\Gamma \leftarrow$ UPDATEPLANS$(\rho_{t-1}, \sigma_{t-1}, \boldsymbol{A_t}, \Gamma)$;
7       **if** IDLE$(\gamma_{t-1}, \sigma_{t-1})$ **then**
8           $ts \leftarrow$ CHOOSEREQUEST$(\gamma_{t-1}, \boldsymbol{A_t}, \Gamma)$;
9           $\sigma_t \leftarrow \sigma_{t-1}$;
10          **for** $i \in 1..m$ **do**
11              $\rho_i \leftarrow \gamma_{t-1}(i)$;
12              **if** IDLE$(\rho_i, \sigma_{t-1}) \wedge ts(i) \neq \perp$ **then**
13                  $\sigma_t \leftarrow \sigma_{t-1}[\text{LAST}(\rho_i) \leftarrow t]$;
14                  $\rho_i \leftarrow \rho_i : ts(i)$;
15                  $\Gamma \leftarrow \{ \gamma \in \Gamma \mid \text{FIRST}(\gamma - \gamma_{t-1})(i) = ts(i) \}$;
16              $\gamma_t \leftarrow \langle \rho_1, \ldots, \rho_m \rangle$;
17          **else**
18              $\gamma_t \leftarrow \gamma_{t-1}$;
19              $\sigma_t \leftarrow \sigma_{t-1}$;
20          $\Gamma \leftarrow \Gamma \cup$ GENERATESOLUTIONS$(\gamma_t, \sigma_t, \boldsymbol{A_t})$;
21   **return** $(\gamma_h, \sigma_h)$;

**Figure 8.7:** The Online Stochastic Routing Algorithm for Multiple Vehicles.

15). When a vehicle is not idle, the decision $ts(i)$ is not processed since the vehicle is not ready yet. This decision may change before the vehicle is ready, either because of new requests or because the generation of new plans identifies a better decision.

The main challenge in multiple vehicle routing is how to make the global decision captured in the tuple $ts$. A naive implementation of CHOOSEREQUEST for algorithm $\mathcal{C}$ is shown in figure 8.8. The implementation initializes the evaluation of all possible tuples (lines 1 and 2), retrieves the next decision tuple from each routing plan $\gamma \in \Gamma$ produced by the optimization algorithm, and increments its evaluation. The tuple with the best evaluation is selected in line 6. Observe that $\gamma - \gamma_t$ denotes the set of route suffixes

$$\gamma - \gamma_t = (\gamma(1) - \gamma_t(1), \ldots, \gamma(m) - \gamma_t(m))$$

and that function FIRST has been lifted from sequences to tuples of sequences, that is,

$$\text{FIRST}(\gamma - \gamma_t) = (\text{FIRST}(\gamma(1) - \gamma_t(1)), \ldots, \text{FIRST}(\gamma(m) - \gamma_t(m))).$$

CHOOSEREQUEST-$\mathcal{C}(\gamma_t, \boldsymbol{A_t}, \Gamma)$
1   **for** $tr \in Customers^m$ **do**
2       $f(tr) \leftarrow 0;$
3   **for** $\gamma \in \Gamma$ **do**
4       $tr = \text{FIRST}(\gamma - \gamma_t);$
5       $f(tr) \leftarrow f(tr) + 1;$
6   **return** arg-max$(tr \in Customers^m) \; f(tr);$

**Figure 8.8:** A Naive Generalization of the Consensus Algorithm for Multiple Decisions.

CHOOSEREQUEST-$\mathcal{C}(\gamma_t, \boldsymbol{A_t}, \Gamma)$
1   **for** $r \in Customers$ **do**
2       $f(r) \leftarrow 0;$
3   **for** $\gamma \in \Gamma$ **do**
4       $tr = \text{FIRST}(\gamma - \gamma_t);$
5       **for** $i \in 1..m$ **do**
6           $f(tr(i)) \leftarrow f(tr(i)) + 1;$
7       $\gamma^c = \text{arg-max}(\gamma \in \Gamma) \sum_{i=1}^{m} f(\text{FIRST}(\gamma - \gamma_t)(i));$
8   **return** $\text{FIRST}(\gamma - \gamma^c);$

**Figure 8.9:** A Pointwise Implementation of the Consensus Algorithm for Multiple Decisions.

Unfortunately, this implementation of consensus is not particularly effective, especially when there are many requests and few samples. Indeed the information about decisions is now distributed over tuples of requests instead of over individual requests. Hence, consensus does not capture the desirability of serving particular requests and evaluates only joint decisions.

This limitation can be remedied by a two-step strategy called *pointwise decisions*. In the first step, each decision is evaluated individually across all scenarios and independently of the vehicles they are allocated to. In the second step, each plan $\gamma \in \Gamma$ is evaluated in terms of its individual decisions and the best plan is selected. Figure 8.9 depicts the pointwise implementation of $\mathcal{C}$. Lines 1 through 6 implement the first step. The implementation initializes the customer evaluations in lines 1 and 2 and considers each plan $\gamma \in \Gamma$. It retrieves the tuple $tr$ of decisions for $\gamma$ (line 4) and increments the evaluation of every request served in $tr$ (lines 5 and 6). The second step takes place in lines 7 and 8 and is particularly interesting. Observe that the implementation cannot select the tuple

$$(r_1^*, \ldots, r_m^*)$$

consisting of the $m$ requests with the $m$ best evaluations $f(r_i^*)$ $(1 \leq i \leq m)$: this tuple may not satisfy the problem-specific constraints or ensure that the service guarantees will be honored. As a result, the implementation considers only the decision tuples coming from the plans in $\Gamma$ since, by construction, these plans satisfy all constraints. Each plan $\gamma \in \Gamma$ is evaluated by summing the evaluations of its next requests $(r_1, \ldots, r_m)$, that is,

$$f(\gamma) = \sum_{i=1}^{m} f(r_i),$$

or, more explicity,

$$f(\gamma) = \sum_{i=1}^{m} f(\text{FIRST}(\gamma - \gamma_t)(i)).$$

The plan $\gamma^c$ with the best evaluation is selected in line 7 and its tuple of decisions is returned in line 8. By ranking the plans according to the individual requests they serve, pointwise decision maintains feasibility while distinguishing the quality of the plans much more precisely than the naive implementation.

## 8.7 Notes and Further Reading

Service guarantees and pointwise decisions were first formalized in [10]. That paper also contains the first formalization of the waiting strategy and its experimental evaluation on vehicle dispatching. Many ideas were in fact already presented as early as 2002 in the online stochastic vehicle-routing algorithm described in [12]. The idea of preserving multiple plans in online vehicle routing was proposed by Taillard et al. [40], but their algorithm does not exploit stochastic information. They maintain multiple plans produced by local search in order to increase the probability of accommodating new requests in the future. Campbell and Savelsbergh [25] use stochastic information to decide whether to accept or reject a request in online grocery problems. Such strategies are evaluated experimentally in chapter 10.

# 9 Online Vehicle Dispatching

*Le génie n'est qu'une plus grande aptitude à la patience.*
— Comte de Buffon

Online vehicle dispatching has numerous industrial applications ranging from long-distance courier to air-taxi services. These applications are challenging both because of their scale and the complexity of the underlying optimization problems. Instances may consider multiple vehicles, hundreds of customers, and a time horizon of eight hours, implying that there is time only for a couple of optimizations between decisions. As a result, algorithm $\mathcal{E}$ is completely impractical, since it cannot even evaluate a tiny portion of the possible decisions on a single scenario.

This chapter studies online vehicle routing on the set of instances described in section 9.1 and derived from [70]. It presents the setting of the algorithms in section 9.2 and reports the experimental results in section 9.3. The experimental results include a comparison between greedy heuristics, local optimization, and the consensus algorithm $\mathcal{C}$ with and without a waiting strategy. Because of the high quality of some of the algorithms, it was not necessary to implement the regret algorithm $\mathcal{R}$, which explains why no such results are reported. This chapter also reports some experimental results on the robustness of algorithm $\mathcal{C}$ and the importance of sophisticated optimization procedures. Finally, we present a visualization of the algorithms, which nicely explains why their performance differs so substantially.

## 9.1   The Online Vehicle Dispatching Problems

The application is based on the instances proposed in [70] to model long-distance courier mail services and it uses a time horizon of eight hours. The instances also have the following properties.

**Customers**   As discussed in chapter 8, some customers are known at the beginning of the day, while others are revealed online as the algorithm proceeds. The instances are generated for various degrees of dynamism (DOD), where the degree of dynamism is the ratio of the number of online customers over the total number of customers. The DODs are taken in the set $\{0\%, 5\%, \ldots, 100\%\}$. For a DOD $x$, there are $n(1-x)$ customers known initially. The remaining customers are generated using an exponential distribution with parameter $\lambda = \frac{nx}{h}$ for their interarrival times. It follows from the associated Poisson distribution (with parameter $\lambda h$) that the expected number of online customers is $nx$, the expected number of customers is $n$, and the expected DOD is $x$. In these instances, the expected number of customers is 160. There are 15 instances for each DOD.

**Customer Locations**   The customers are distributed in a 20km×20km region under two models: M3 and M4. In model M3, the customers are generated using a uniform distribution while, in

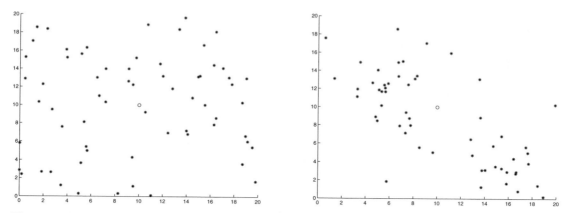

**Figure 9.1:** Two Instances of Models M3 and M4 for Customer Locations.

model M4, they are generated from 2-dimensional Gaussians centered at two points in the region. Examples of the uniform and Gaussian distributions are seen in figure 9.1.

**Service Times**  The service times for the customers are generated according to a log-normal distribution with parameters (.8777, .6647). The mean service time is three minutes and the variance is five minutes. These service times were chosen in [70] to mimic the service times of long-distance courier mail services.

**Vehicles**  There are four vehicles, each of which has a uniform speed of 40 km/h. This means that minimizing travel distances and travel times are equivalent. Each vehicle can serve at most 50 customers and the vehicle must return to its depot by the time horizon.

## 9.2  Setting of the Algorithms

The online algorithms are implemented with the following settings.

**Initial Plans**  Before starting the online process, 25 different scenarios are generated and optimized for 1 minute using LNS. These initial solutions are used to determine the first customer of each vehicle. An additional 25 scenarios are then generated and optimized for another minute, since this was shown to improve the overall quality of the results. This requires about 50 minutes of computation time before the time horizon H (for instance, at the beginning of the day), which is reasonable for this type of application. Subsequent scenarios are "optimized" with large neighborhood search for about 10 seconds.

**LNS Parameters**  The parameters for LNS are as follows: 30 for the maximum number of customers to relocate ($k = 30$), 100 attempts at relocating $i$ customers without improvement before trying $i + 1$ customers (*maxIterations=100*), 15 for the determinism factor ($\theta = 15$), and 4 discrepancies.

**Accepting or Rejecting Requests**  A simple insertion heuristic is used to decide whether to accept a new request: a request is accepted if it can be inserted into a routing plan and rejected otherwise.

## 9.3  Experimental Results

This section presents the experimental results for online vehicle dispatching. The results compare the online stochastic algorithms with the nearest-neighbor heuristic from [70] and a local optimization algorithm inspired by [40]. This section also presents robustness results of the online stochastic algorithms and discusses, once again, the importance of a sophisticated optimization method in obtaining high-quality solutions. It also discusses the set $\Gamma$ of plans that varies in size over time and the anticipativity assumption.

### 9.3.1  The Algorithms

Two oblivious algorithms are used for comparison purposes: a nearest-neighbor heuristic (NN) and a multiplan local optimization method (LO).

- The NN heuristic was shown very effective for single vehicle dispatching in [70]. It was generalized for multiple vehicles and to provide service guarantees. In particular, whenever a request arrives, the NN algorithm is simulated to determine whether it can accommodate the new request, in which case the request is accepted. Otherwise, the request is rejected.

- The LO algorithm is an abstraction of the method proposed by [40]. Its basic idea is to keep multiple plans that are optimized by a local search algorithm on the *known* customers. The best available plan is used at time $t$ to decide how to serve a request.

These oblivious algorithms are compared with the consensus algorithm without and with a waiting strategy ($\mathcal{C}$ and $\mathcal{CW}$). As mentioned earlier, the regret algorithm was not implemented for vehicle dispatching given the excellent quality of consensus. Note also that the online algorithms are also compared with the best offline solutions found using LNS. This is not the true optimum solution, which is extremely difficult to obtain on multiple vehicle dispatching, but it still represents the "best" solution that the online algorithms could achieve when using LNS as the optimization black-box. The results were run on an AMD Athlon 64 3000 processor with 512 MB of RAM under the Linux operating system.

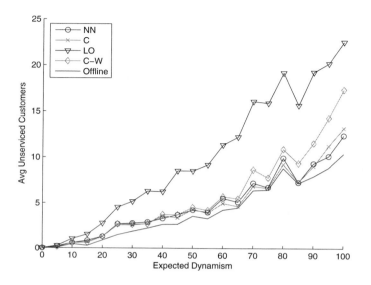

**Figure 9.2:** Results on the Number of Serviced Customers for Model M3.

### 9.3.2 Customer Service

Figures 9.2 and 9.3 depict the experimental results on the number of missed customers for models M3 and M4 under various degrees of dynamism. The results indicate clearly that algorithm LO is dominated by other approaches. Its pathological behavior, which is demonstrated visually in section 9.4, comes from the over-optimization of the travel times due to the absence of information on future requests. The remaining approaches service a comparable number of customers, except for very high degrees of dynamism for which algorithm $\mathcal{CW}$ is slightly dominated by NN and $\mathcal{C}$. Observe that stochastic information does not bring significant benefits as far as customer service is concerned on vehicle dispatching.[1]

### 9.3.3 Travel Distances

Figures 9.4 and 9.5 depict the results for travel distances on models M3 and M4. No results are shown for algorithm LO, which is not competitive as far as customer service is concerned. Recall also that travel distances and travel times are directly proportional in these problems.

The experimental results are rather interesting as they indicate that the stochastic instantiations

---

[1] This contrasts with the online vehicle routing applications presented in chapter 10.

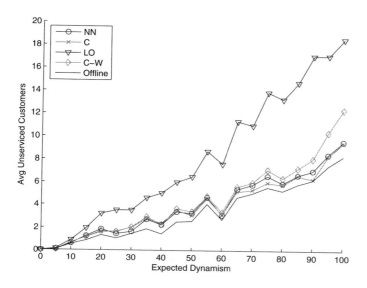

**Figure 9.3:** Results on the Number of Serviced Customers for Model M4.

of the online algorithm significantly reduce travel distances when compared to NN. The results are particularly impressive for algorithm $\mathcal{CW}$, which bridges almost the entire gap between the NN heuristic and the offline optimum. The waiting strategy is particularly effective in reducing travel distances because $\mathcal{CW}$ significantly improves upon $\mathcal{C}$, which itself clearly dominates NN. It is also very satisfying to observe that the travel distances are essentially not affected by the degree of dynamism for algorithm $\mathcal{CW}$, indicating that the algorithm anticipates the future effectively. Indeed all instances depicted in the results have the same number of expected customers (160). What differs is how many of them are known before the algorithm starts. The experimental results show that $\mathcal{CW}$ is not sensitive to this degree of dynamism (except for very high DODs).

### 9.3.4   Robustness with Respect to Noisy Distributions

It is natural to wonder how the algorithms behave when the stochastic information is noisy. This situation could arise from faulty historical data, predictions, and/or approximations in machine learning algorithms. Figure 9.6 depicts some results for model M3 and DODs of 20% and 50%, when the stochastic information is noisy. The noise in these results consists of giving the algorithms a distribution with a different number of online customers. More precisely, the algorithms are given distributions predicting the number of customers fewer or more customers than the expected 32 and

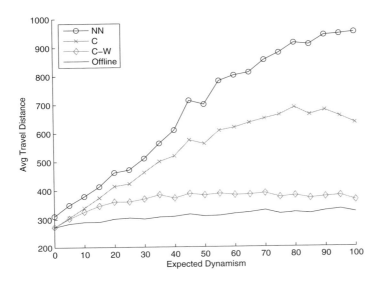

**Figure 9.4:** Experimental Results on Travel Distance for Model M3.

80 online customers, and the figure depicts their performance under these noisy conditions. The figure shows both the effect of noisy distributions on the number of served customers (left graphs) and on travel distances (right graphs). The x-axis depicts the number of online requests predicted by the noisy distributions, which differ from the 32 customers expected in the top graphs and the 80 customers expected in the bottom graphs.

The results seem to indicate that it is better to be optimistic when estimating the number of dynamic customers. For example, on 20% dynamism, algorithm $\mathcal{CW}$ serves roughly the same number of customers when the noisy distribution generates between 20 and 100 dynamic customers. However, it performs the best in terms of travel distance when the noisy distribution predicts 50 dynamic customers, slightly more than the 32 expected online requests. The performance degradation mostly occurs when the noisy distribution predicts very few online customers. This is not surprising, since algorithm $\mathcal{C}$ and $\mathcal{CW}$ then degenerate into a form of local optimization, which was shown to perform poorly.

### 9.3.5 The Importance of Optimization

Section 6.3.5 discussed the tradeoff between sampling and optimization for online multiknapsacks and indicated that, whenever enough time is available, it is desirable to use a more sophisticated

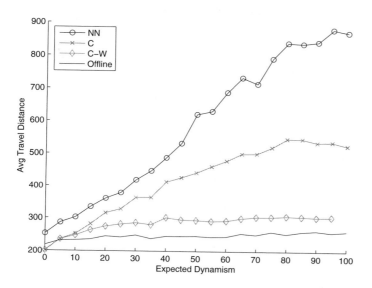

**Figure 9.5:** Experimental Results on Travel Distance for Model M4.

optimization procedure. This section reconsiders this tradeoff on online vehicle dispatching. This is particularly noteworthy in this context, since the algorithms were implemented to run under operational conditions: they are event-driven and interrupted as soon as a decision must take place. Figures 9.7, 9.8, 9.9, and 9.10 compare algorithm $\mathcal{C}$, which uses LNS, with a version of the consensus algorithm using heuristic NN to "solve" the scenarios ($\mathcal{C}$(NN) in the figures). Algorithm $\mathcal{C}$(NN) considers many more scenarios than $\mathcal{C}$ but returns solutions of lower quality. The figures give results both for the number of served customers and travel distances. Once again, there is not much difference for the number of served customers (except for very high degrees of dynamism). However, the results on travel distances are particularly interesting: they indicate that the use of LNS, a sophisticated optimization procedure, brings significant benefits over NN.

## 9.3.6 Routing Plans

Figures 9.11 and 9.12 depict the number of routing plans over time for algorithms $\mathcal{C}$ and $\mathcal{CW}$ on a model M4 instance with a 60% DOD. The results highlight significant differences between the algorithms. Algorithm $\mathcal{C}$ typically has fewer than 5 plans and never more than 10. In contrast, algorithm $\mathcal{CW}$ rarely has fewer than 5 plans, has many peaks with more than 10 plans, and can have up to 35 plans. This difference is explained by the waiting strategy: algorithm $\mathcal{CW}$ often waits

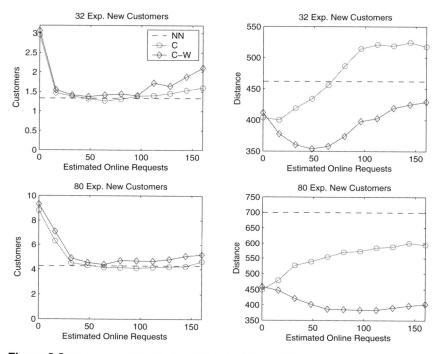

**Figure 9.6:** Robustness Results for $\mathcal{CW}$ under Model M3.

before committing to a customer and can accumulate many plans during these waiting periods. When $\mathcal{CW}$ finally decides to serve some customers, the number of plans experiences a significant drop but remains much higher than in $\mathcal{C}$. Note that, as should be clear from the packet scheduling and multiknapsack problems, more plans typically mean more-informed decisions, which is another reason for the excellent performance of $\mathcal{CW}$. Indeed a side effect of the waiting strategy is the ability to make decisions using a large number of samples.

### 9.3.7 The Anticipativity Assumption

Figure 9.13 studies the anticipativity assumption on vehicle dispatching for algorithm $\mathcal{CW}$. At each decision point, the best tuple of decisions for a scenario $\gamma$ is compared with the selected tuple of decisions. There is a match when the same customer is served next on the same vehicle and each scenario is attributed a matching percentage. The figure reports the average matching percentage. In general, the matching percentages are high and oscillate between 60% and 80%. There are some peaks above 80% (mostly late in the day) and some valleys below 60% (mostly earlier in the day).

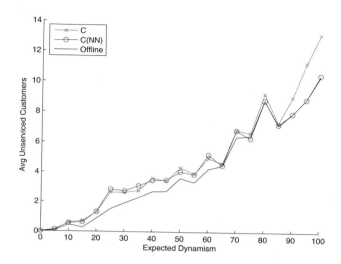

**Figure 9.7:** Results on the Number of Serviced Customers for M3.

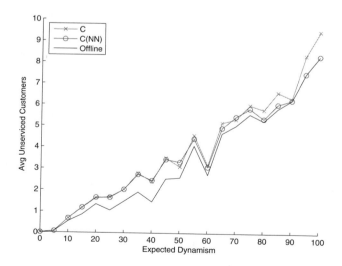

**Figure 9.8:** Results on the Number of Serviced Customers for M4.

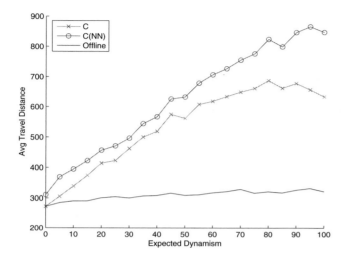

**Figure 9.9:** Results on Travel Distance for M3.

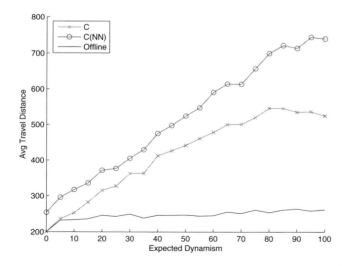

**Figure 9.10:** Results on Travel Distance for M4.

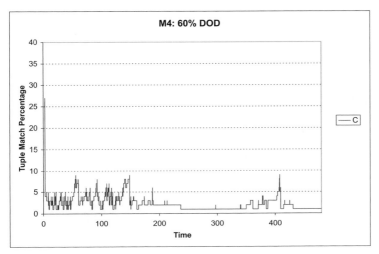

**Figure 9.11:** The Number of Available Routing Plans over Time: Algorithm $\mathcal{C}$.

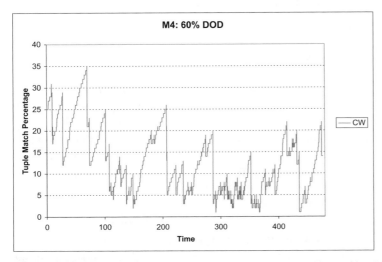

**Figure 9.12:** The Number of Available Routing Plans over Time: Algorithm $\mathcal{CW}$.

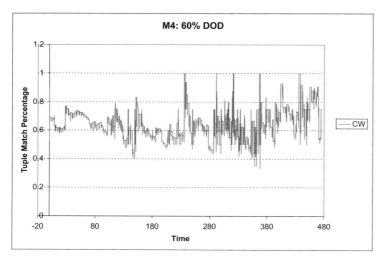

**Figure 9.13:** The Anticipativity Assumption in Vehicle Dispatching: Model M4.

These variations typically coincide with the increases or drops in the number of plans as depicted in figure 9.12. The anticipativity assumption seems largely valid once again in this application, which is not suprising given the small distance between the offline, a posteriori optimum and the online solution produced by algorithm $\mathcal{CW}$.

## 9.4   Visualizations of the Algorithms

This section presents a visualization of the algorithms over time. The goal is to explain the experimental results intuitively in terms of the actual decisions taken by the algorithms and to provide insights on their differences in behavior and solution quality. The visualizations consider only one run of the algorithms (with 50% DOD), although other runs typically exhibit similar behaviors. They report three snapshots for each algorithm, depicting the customers visited after 1 hour, 4 hours, and 8 hours respectively. Each snapshot shows the four vehicles, one in each quadrant. The accepted customers at the time of the snapshot are shown in red (light gray) and those who are served by the algorithm are shown in dark. All accepted and rejected customers are shown in all quadrants, since it is not clear which vehicles will actually serve them. Note that the right side of each snapshot will provide some useful information. It depicts the expected number of customers, the degree of dynamism, the number of plans available at this stage, the number of unserviced and rejected customers, and the travel costs. The projected travel cost of each available plan is also shown.

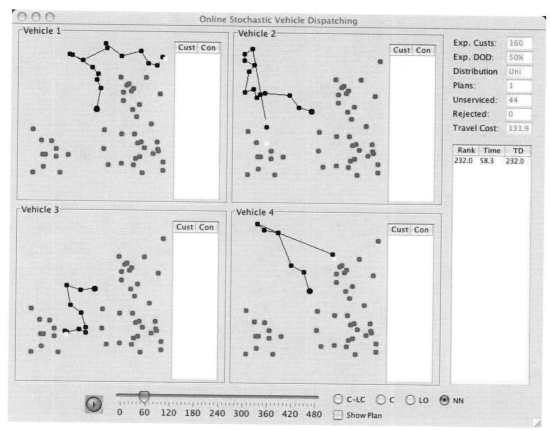

**Figure 9.14:** Algorithm NN after 1 Hour.

## 9.4.1   The Nearest Neighbor Algorithm

Figures 9.14, 9.15, and 9.16 visualize algorithm NN. After 1 hour, algorithm NN has traveled 131.9km and expects to travel 232km. It still has 44 unserviced customers and has not rejected any requests. After 4 hours, algorithm NN has traveled 403.8km and has visited all the known customers. After 8 hours, algorithm NN has rejected 3 customers and traveled 580.7km. It is important to point out two fundamental aspects in NN's behavior:

1. After 4 hours, algorithm NN becomes essentially a first-come/first-serve algorithm. All known customers have been served at that point and NN simply dispatches the closest vehicle for each incoming request. This observation is revisited when algorithm $\mathcal{C}$ is discussed.

2. Each vehicle visits a significant region of the space and its travel pattern exhibits many cross-

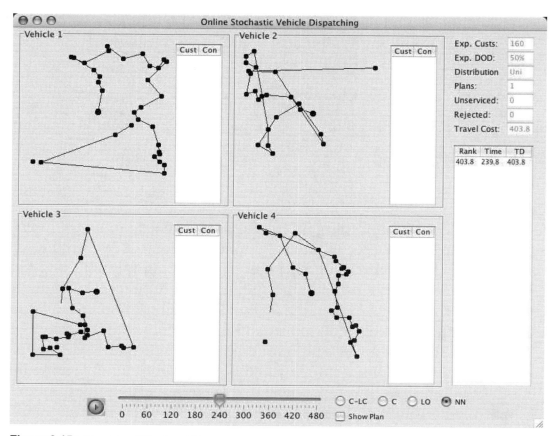

**Figure 9.15:** Algorithm NN after 4 Hours.

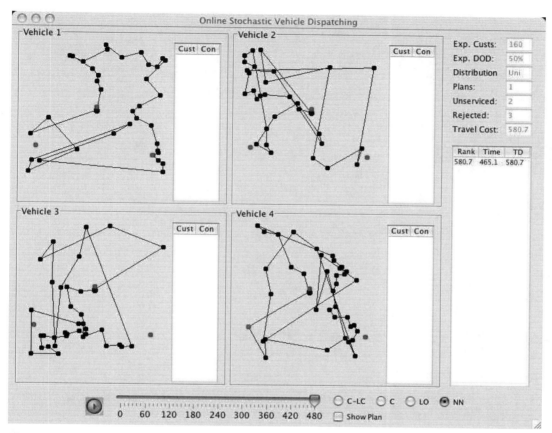

**Figure 9.16:** Algorithm NN after 8 Hours.

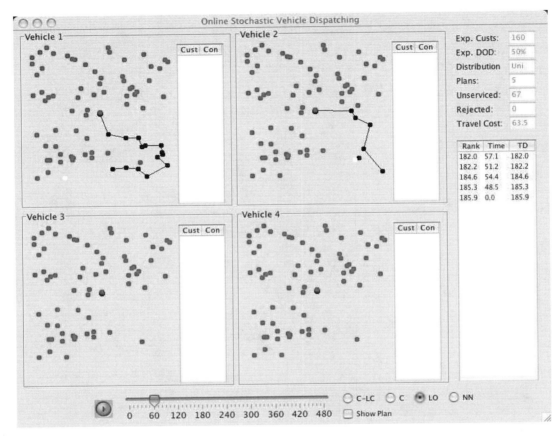

**Figure 9.17:** Algorithm LO after 1 Hour.

ings. Whether this can be avoided is not clear at this point since the customers arrive at different times of the day. Once again, this observation is revisited for algorithm $\mathcal{C}$.

### 9.4.2 The Local Optimization Algorithm

Figures 9.17, 9.18, and 9.19 visualize algorithm LO, whose behavior is particularly noteworthy. After 1 hour, algorithm LO has traveled 63.5km and expects to travel 182km in its best plan. It still has 67 unserviced customers and has not rejected any request. In other words, it has traveled less than algorithm NN and anticipates a smaller total travel time. Moreover, it has deployed only two vehicles at this stage, since using fewer vehicles typically means shorter travel times in these problems. Algorithm LO also has five available plans for the accepted customers. After 4 hours,

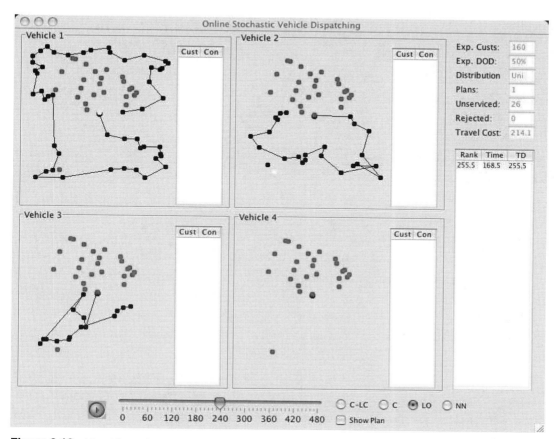

**Figure 9.18:** Algorithm LO after 4 Hours.

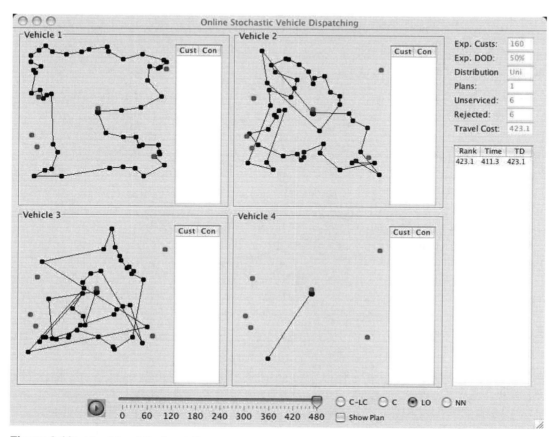

**Figure 9.19:** Algorithm LO after 8 Hours.

algorithm LO has traveled 214.1km, deployed three vehicles, still has 26 customers to serve, and predicts a travel distance of 255.5km. It has only one plan for accepted customers at this stage. After 8 hours, algorithm LO has rejected 6 customers, traveled 423.1km, and finally deployed its last vehicle to serve only one customer.

The visualization explains why algorithm LO is dominated by NN as far as customer service is concerned. Indeed algorithm LO over-optimizes travel distances, refraining to deploy all vehicles until late in the day since deploying fewer vehicles typically means shorter travel distances. As a consequence, algorithm LO leaves little room to accommodate new requests at the end of the day because the vehicles are badly positioned to serve them quickly.

### 9.4.3   The Consensus Algorithm

Figures 9.20, 9.21, and 9.22 visualize algorithm $\mathcal{C}$. After 1 hour, algorithm $\mathcal{C}$ has traveled 121.9km and expects to travel 273.3km. It still has 47 unserviced customers and has not rejected any request. After 4 hours, algorithm $\mathcal{C}$ has traveled 348.2km and has visited all the known customers but one. After 8 hours, algorithm $\mathcal{C}$ has rejected 3 customers and traveled 518.2km. It is also useful to point out several fundamental aspects of algorithm $\mathcal{C}$:

1. Algorithm $\mathcal{C}$ exhibits "nice" travel patterns after 4 hours with good geographic concentration and relatively few crossings. However, the travel patterns deteriorate in the second half of the day, as the vehicles travel to relatively remote customers and exhibit some significant crossings.

2. Algorithm $\mathcal{C}$ has very few routing plans available. It has only two plans after 1 hour and a single plan after 4 hours.

3. All accepted customers have been served after 4 hours. As a result, $\mathcal{C}$ also becomes a first-come/first-serve algorithm after 4 hours, explaining why its travel patterns deteriorate so much. In a sense, algorithm $\mathcal{C}$ has been too eager to serve the accepted requests and cannot amortize some of its travel with future requests.

### 9.4.4   The Consensus Algorithm with a Waiting Strategy

Figures 9.23, 9.24, and 9.25 visualize algorithm $\mathcal{CW}$.

- After 1 hour, algorithm $\mathcal{CW}$ has traveled only 21.6km and expects to travel around 237km. It still has 83 unserviced customers, has not rejected any request, and has a wealth of routing plans to assist in the decision-making process. Algorithm $\mathcal{CW}$ deploys the vehicles very slowly but, contrary to algorithm LO, deploys them all early because it can anticipate the expected number of online customers.

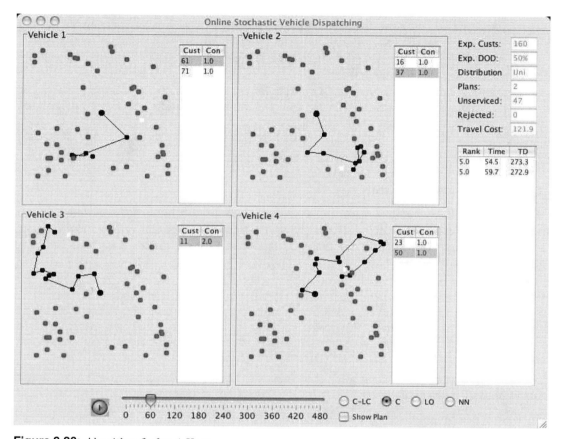

**Figure 9.20:** Algorithm $\mathcal{C}$ after 1 Hour.

- After 4 hours, algorithm $\mathcal{CW}$ has traveled only 150.4km, has 62 unserviced customers, predicts a travel distance around 283km, and still has 8 routing plans available for the decision process. Recall that algorithm $\mathcal{C}$ has served all accepted customers (but one) after 4 hours. This strongly contrasts with algorithm $\mathcal{CW}$, which still has 62 unserviced customers.

- After 8 hours, algorithm $\mathcal{CW}$ has rejected 3 customers and traveled only 363.9km.

There are a few significant features of $\mathcal{CW}$ that deserve to be emphasized.

1. The resulting routing plan is quite appealing visually. The vehicles serve distinct regions of the space and exhibit few crossings. This is an emerging behavior: algorithm $\mathcal{CW}$ has no understanding of geographic locations other than through its optimization algorithm.

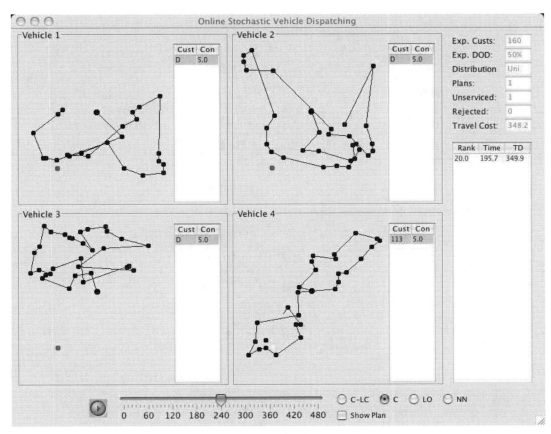

**Figure 9.21:** Algorithm $\mathcal{C}$ after 4 Hours.

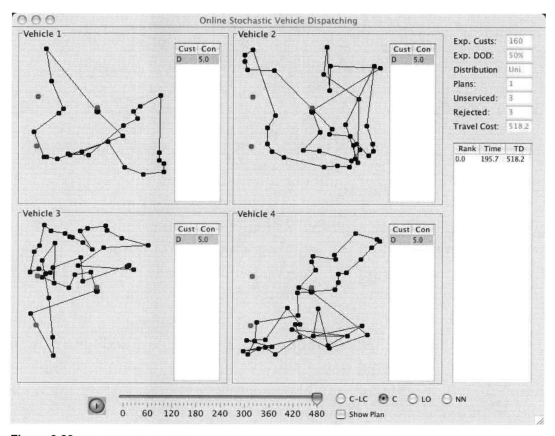

**Figure 9.22:** Algorithm $\mathcal{C}$ after 8 Hours.

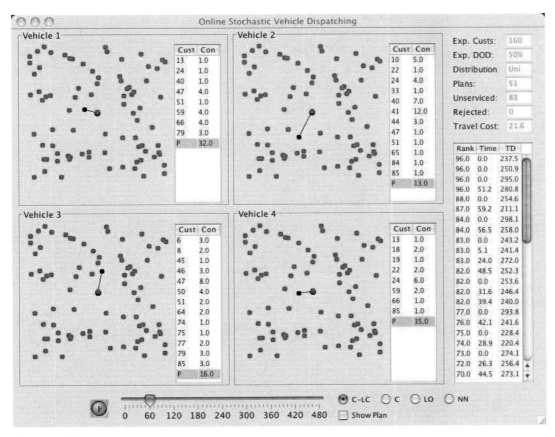

**Figure 9.23:** Algorithm $\mathcal{CW}$ after 1 Hour.

2. The waiting strategy in $\mathcal{CW}$ is beneficial only in an online setting and is not useful for pure offline problems. Moreover, the waiting strategy can be built naturally on top of a traditional offline algorithm, leveraging traditional optimization technology for a new class of applications.

3. It is the synergy between a sophisticated optimization procedure, stochastic information, the consensus algorithm, and the waiting strategy that accounts for the high quality of the plans. Each component is critical and plays an orthogonal role in achieving the results.

## 9.5   Notes and Further Reading

The use of historical data [39, 40, 61] or probabilistic models in [43, 115] in vehicle routing has been advocated in many other papers. The online vehicle-dispatching instances studied in this

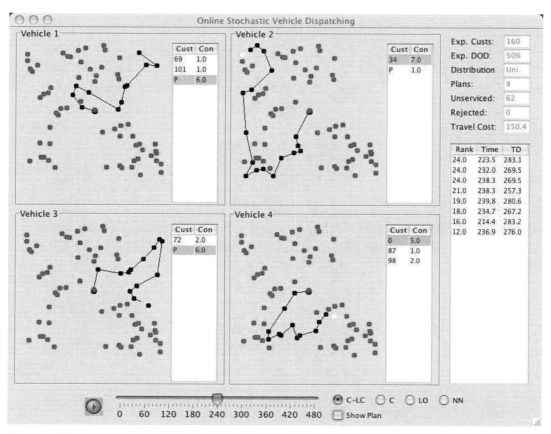

**Figure 9.24:** Algorithm $\mathcal{CW}$ after 4 Hours.

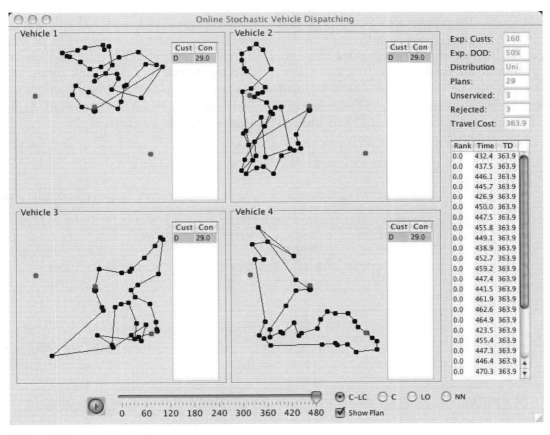

**Figure 9.25:** Algorithm $\mathcal{CW}$ after 8 Hours.

chapter generalize the problems studied in [70], which involve a single vehicle only and no capacity constraints. Preliminary results on these problems studied in this chapter are reported in [8, 10]. Recently, waiting heuristics have attracted more attention (see, for instance, [78, 79]). The beauty in the algorithms presented here is that the choice of when and where to wait is fully automatic and systematically derived from stochastic information. Indeed, the waiting strategy is a natural instantiation of our generic online stochastic framework. This contrasts with heuristics to distribute waiting time in routing plans (e.g., [78, 79]) which do not use stochastic information.

# 10 Online Vehicle Routing with Time Windows

*If you torture the data enough, nature will always confess.*
— Ronald Coase

This chapter presents experimental results for online multiple vehicle routing with time windows. Contrary to the vehicle-dispatching applications discussed in chapter 9, the main difficulty in the vehicle-routing applications is to serve as many customers as possible. The online instances are derived from the Solomon benchmarks, a collection of very challenging routing problems studied by many authors. Their complexity stems from the time windows that restrict the times when customers can be served and hence impose some strong constraints on the routing plans.

This chapter compares all algorithms on the online vehicle-routing applications including the consensus and regret algorithms and their variations with waiting and relocation strategies. The algorithms are compared on instances with various degrees of dynamism and with different structures for the request times of the customers, giving insights on the benefits and limitations of the various approaches. The experimental results are particularly illuminating and emphasizes the significance of the relocation strategy in online vehicle routing with time windows. This is especially remarkable given the simplicity of the relocation strategy that only exploits stochastic information and does not use any problem-specific information. Its insight is simply to select vehicle movements using both accepted and sampled requests.

The rest of this chapter is organized as follows. Section 10.1 describes the online instances and section 10.2 presents the experimental setting. Section 10.3 describes the experimental results for the number of served customers. It also presents the number of plans, the anticipativity assumption, the waiting times of the customers, and the evaluation of strategies to accept or reject customers using stochastic information.

## 10.1 The Online Instances

The online vehicle-routing problems are generated from the Solomon benchmarks [105], a collection of very challenging multiple vehicle-routing problems with time windows and 100 customers. Each customer from the Solomon instances becomes a customer region in the online problems. The expected number of customer requests in the online instances is 100 and the expected number of requests from a particular customer region is one. As a result, the online instances preserve much of the structure of the Solomon instances.

**Request Arrivals** To specify the arrival distribution of customer requests, the time horizon $H = l_0 - e_0$ is divided into four time periods. Period 0 corresponds to known requests, that is, requests available before the start of the day. Periods 1 to 3 can be thought of as morning, early afternoon,

| Customers | | | | | | |
|---|---|---|---|---|---|---|
| | Type 1 | Type 2 | | Type 3 | | |
| Period | 0 | 0 | 1 | 0 | 1 | 2 |
| Class 1 | 1.00 | 0.50 | 0.50 | 0.50 | 0.40 | 0.10 |
| Class 2 | 1.00 | 0.50 | 0.50 | 0.50 | 0.10 | 0.40 |
| Class 3 | 1.00 | 0.50 | 0.50 | 0.50 | 0.25 | 0.25 |
| Class 4 | 1.00 | 0.50 | 0.50 | 0.20 | 0.20 | 0.60 |
| Class 5 | 1.00 | 0.10 | 0.90 | 0.10 | 0.10 | 0.80 |

**Table 10.1:** Online Vehicle Routing: Distribution of the Request Times.

and late afternoon. Requests arrive only in periods 0 through 2 to ensure that the offline problems can serve all customers.

Customer regions are partitioned into three types according to their time windows (that is, when they can be served) and their distances from and to the depots. Once again, the distances from and to the depot are taken into account to generate online instances whose offline counterparts can serve all requests.

Type-1 customers request service in period 0, which means that their requests are known at the beginning of the day. Type-2 customers request service in periods 0 or 1 according to some distribution. Type-3 customers place their requests in the first three periods, once again according to some distribution.

The instances are divided into five classes according to their probability distributions. The probabilities for the five classes are specified in table 10.1. For example, a type-2 customer in class 1 makes a request in period 0 with probability 0.50 and in period 1 with the same probability. A type-3 customer in class 1 makes a request in period 0 with probability 0.50, in period 1 with probability 0.40, and in period 2 with probability 0.10. These probabilities are independent and allow multiple or no requests from a particular region. Observe that the probabilities of late requests in classes 4 and 5 are significantly larger, making these instances more challenging.

The request arrival times were generated as follows. If there is a request in time period $k$, then the arrival time for a customer $c$ is drawn uniformly at random from the time interval

$$[(k-1) * \frac{h}{3}, \min(\lambda_c, k * \frac{h}{3} - 1)]$$

where $\lambda_c$ is the latest time a vehicle can depart from its depot, service customer $c$, and return to its depot.

**Customer Data**   The customer data (that is, the customer locations, time windows, demands, and service times) were taken directly from the class RC in the Solomon benchmarks. RC problems

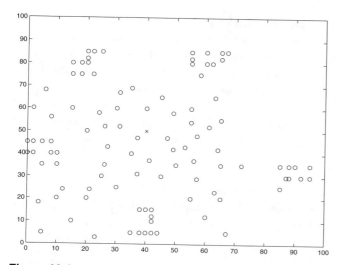

**Figure 10.1:** The Customer Region Locations in the RC Class of the Solomon Benchmarks.

have a mix of clustered and randomly distributed customers, which makes them particularly interesting. Figure 10.1 shows the location of the customer regions in the RC class of the Solomon benchmarks. Moreover, the problems were selected with diverse time windows. In particular, RC101 problems have a good mix of short and long time windows and have a high percentage of type-2 customers. RC102 problems have many regions with short time windows, 45 type-2 customers, and 45 type-3 customers. RC104 problems have regions with extremely long time windows and 83 type-3 customers. As a result, the instances cover a wide spectrum of online instances.

**Number of Vehicles**  For each instance, the number of vehicles available for the online algorithms was determined by solving the offline, a posteriori optimization problem and adding two vehicles. The offline problems were solved through the two-stage hybrid algorithm in [9], a particularly effective algorithm to minimize the number of vehicles.

## 10.2  Experimental Setting

All algorithms are executed on an AMD Athlon 64 3000 processor with 512MB of RAM running Linux. Each of the instances is run fifty times to account for the nondeterministic nature of the algorithms.

**The Algorithms**   The results compare local optimization with the consensus and regret algorithms, which may include the waiting and relocation strategies. Algorithm LO is a generalization of the parallel tabu-search algorithm in [40]. It generates multiple routing plans using LNS on the accepted customers. These plans are then used to accept or reject new customers and to select the decisions at each time step. LO is thus close to algorithm $\mathcal{C}$, the main difference being that no stochastic information is exploited. The consensus algorithms $\mathcal{C}$, $\mathcal{CW}$, and $\mathcal{CR}$ are direct implementation of the generic online routing algorithms. The regret algorithms $\mathcal{R}$, $\mathcal{RW}$, and $\mathcal{RR}$ are instantiations of the generic algorithms for a simple and fast suboptimality approximation, which we now describe.

**The Suboptimality Approximation**   Consider the decision of choosing which customer to serve next on vehicle $v$ and let $s_t$ be the first customer on the route of vehicle $v$ at time $t$. To evaluate the regret of another customer $r$ on the same vehicle $v$, the suboptimality approximation determines whether there is a feasible swap of $r$ and $s_t$ on $v$, in which case the regret is zero. Otherwise, if such a swap violates the time-window constraints, the regret is 1.

**Initial Plans and Online Processes**   The online stochastic algorithms generate and solve fifty scenarios to select the decisions at time 0. The algorithms also generate and solve fifty additional samples to create plans given the set of decisions at time 0. Each optimization in this initialization step is allocated one minute. During the online execution, fifty seconds are available between each decision and the online algorithm performs five optimizations, each running for ten seconds.

## 10.3   Experimental Results

We now present the experimental results for online vehicle routing with time windows. Contrary to the vehicle dispatching in chapter 9, the results focus on the number of served customers, which places a significant challenge on these instances.

### 10.3.1   Customer Service

Before discussing the results, it is important to recall that all customers can be served in the offline, a posteriori problem. Moreover, although the results are also averages over fifty runs for each instance, we often omit the words "in the average" for simplicity.

Figure 10.2 summarizes all the results for the regret and consensus algorithms. The graphs depict the various instance results and a linear regression for each class of algorithms. The main result in this chapter is the outstanding behavior of $\mathcal{RR}$. From the interpolations, it can be seen that algorithm $\mathcal{RR}$ dominates all the algorithms for DODs of more than 50 percent and that the improvements increase substantially as the DOD grows. The improvement over $\mathcal{R}$ is also significant and

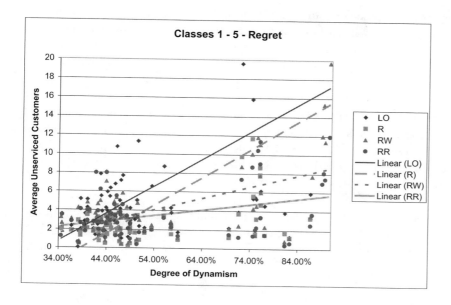

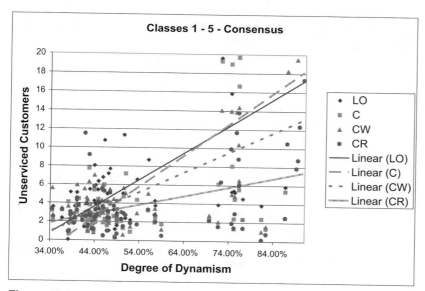

**Figure 10.2:** Online Vehicle Routing: Unserviced Customers for the Regret and Consensus Algorithms.

similar in scope, indicating the importance of using relocation on this set of instances. Subsequent results will characterize more precisely when the relocation strategy is of paramount importance. Algorithm $\mathcal{RW}$ also is quite effective in general, but it is dominated by $\mathcal{RR}$ as the DOD increases. Similar results can be observed for the consensus algorithms $\mathcal{C}$, $\mathcal{CW}$, and $\mathcal{CR}$, but, in general, they serve fewer customers than their regret counterparts.

**Results on Class 1**    Figure 10.3 depicts the results for class-1 instances. On class 1, algorithm $\mathcal{R}$ is the most effective, but $\mathcal{RW}$ and $\mathcal{RR}$ are relatively close in quality. Algorithm $\mathcal{R}$ never misses more than 4.4 customers: this contrasts with LO, which fails to serve as many as 11 customers. The performance of the online stochastic algorithms is excellent in these instances, indicating that stochastic information improves customer service significantly.

**Results on Classes 2 and 3**    Figure 10.4 summarizes the results for classes 2 and 3, whose instances seem easier than other classes. No algorithm loses too many customers on these instances. Once again, algorithm $\mathcal{R}$ is most effective, although all algorithms are reasonably close on these classes.

**Results on Class 4**    Figure 10.5 depicts the results on class 4, which are particularly noteworthy. Algorithms $\mathcal{RR}$ and $\mathcal{RW}$ are reasonably close. What is interesting is the significant gain they yield over LO, $\mathcal{C}$, and $\mathcal{R}$. These results show the significance of the waiting and relocation strategies on the class-4 instances. Observe that LO misses more than twenty five customers on instance rc104-2, while only about ten customers are not served by algorithm $\mathcal{RR}$. On class 4, and particularly on the instances with high DODs, the gain produced by algorithms $\mathcal{RR}$ and $\mathcal{RW}$ is rather impressive. Their consensus counterparts also behave well, which is to be expected since consensus is a form of regret algorithms on this problem.

**Results on Class 5**    Class 5 contains the most difficult instances and the experimental results are depicted in figure 10.6. On these instances, some algorithms may miss up to thirty customers. This is the case of algorithms LO, $\mathcal{C}$, and $\mathcal{R}$ on rc104-4. Once again, algorithm $\mathcal{RR}$ is the most effective algorithm and misses only seven customers on rc104-4. The improvements of $\mathcal{RR}$ over other algorithms is quite dramatic on class 5 and grow with the degree of dynamism. They demonstrate clearly the value of the regret algorithm with a relocation strategy on online vehicle routing.

It is important to point out that the relocation is completely guided by the stochastic information and does not include any problem-specific knowledge: a vehicle simply moves to the selected customer whether this is an accepted or a sampled customer. This contrasts with other approaches (e.g., [71, 111]) where a priori information on the problems, instances, and distributions is heuristically used to select relocation points. As a consequence, the relocation strategy is simple to implement,

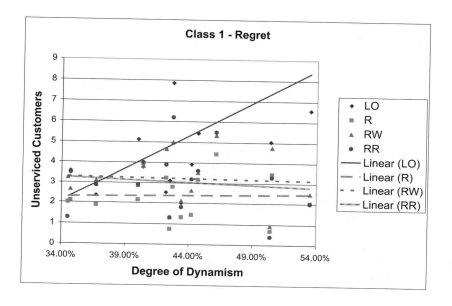

| Problem | DOD | Vehicles | LO | $\mathcal{C}$ | $\mathcal{CW}$ | $\mathcal{CR}$ | $\mathcal{R}$ | $\mathcal{RW}$ | $\mathcal{RR}$ |
|---------|-----|----------|-----|------|------|------|------|------|------|
| rc101-1 | 36.70% | 16 | 2.36 | 3.24 | 3.42 | 3.26 | 1.88 | 3.12 | 2.86 |
| rc101-2 | 34.50% | 16 | 2.00 | 3.16 | 5.62 | 3.31 | 2.02 | 3.26 | 1.28 |
| rc101-3 | 40.40% | 15 | 3.78 | 4.46 | 3.72 | 3.80 | 3.94 | 3.82 | 4.00 |
| rc101-4 | 42.20% | 17 | 2.52 | 2.62 | 4.56 | 3.90 | 3.26 | 4.66 | 3.90 |
| rc101-5 | 34.70% | 17 | 3.58 | 2.06 | 2.68 | 2.50 | 2.10 | 2.66 | 3.50 |
| rc102-1 | 43.40% | 14 | 2.00 | 1.60 | 2.24 | 1.20 | 1.32 | 2.10 | 1.80 |
| rc102-2 | 44.20% | 13 | 3.92 | 2.36 | 3.28 | 2.50 | 1.44 | 2.62 | 3.20 |
| rc102-3 | 40.00% | 15 | 5.12 | 2.98 | 3.34 | 3.20 | 2.14 | 2.86 | 2.90 |
| rc102-4 | 42.50% | 14 | 3.12 | 0.92 | 1.02 | 1.60 | 0.76 | 1.32 | 1.30 |
| rc102-5 | 44.70% | 15 | 5.44 | 2.76 | 3.70 | 2.75 | 3.18 | 3.52 | 3.62 |
| rc104-1 | 46.10% | 11 | 10.76 | 5.98 | 5.40 | 5.30 | 4.44 | 5.34 | 5.50 |
| rc104-2 | 42.70% | 12 | 7.86 | 7.74 | 6.30 | 9.20 | 2.80 | 5.00 | 6.20 |
| rc104-3 | 50.50% | 13 | 11.34 | 4.72 | 5.64 | 2.90 | 3.46 | 4.76 | 3.30 |
| rc104-4 | 50.40% | 12 | 5.04 | 1.32 | 0.62 | 1.50 | 0.70 | 0.94 | 0.40 |
| rc104-5 | 53.50% | 11 | 6.56 | 4.36 | 2.92 | 1.70 | 2.02 | 2.48 | 2.00 |

**Figure 10.3:** Online Vehicle Routing: Unserviced Customers on Class 1.

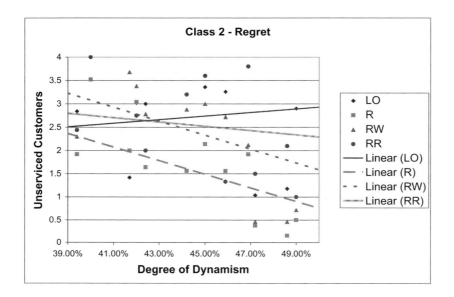

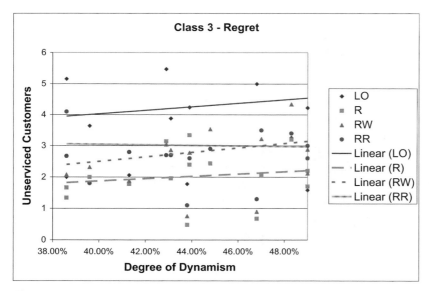

**Figure 10.4:** Online Vehicle Routing: Unserviced Customers on Classes 2 and 3.

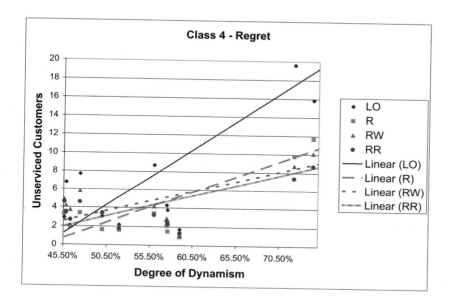

| Problem | DOD | Vehicles | LO | $\mathcal{C}$ | $\mathcal{CW}$ | $\mathcal{CR}$ | $\mathcal{R}$ | $\mathcal{RW}$ | $\mathcal{RR}$ |
|---------|-----|----------|-----|------|------|------|------|------|------|
| rc101-1 | 46.30% | 16 | 2.08 | 2.24 | 4.16 | 3.30 | 1.94 | 3.72 | 2.68 |
| rc101-2 | 45.80% | 15 | 6.78 | 5.42 | 5.94 | 3.62 | 3.50 | 4.18 | 3.44 |
| rc101-3 | 50.00% | 16 | 3.06 | 2.06 | 3.06 | 2.28 | 1.66 | 3.46 | 3.42 |
| rc101-4 | 45.60% | 17 | 2.90 | 3.16 | 4.30 | 5.54 | 3.28 | 4.86 | 4.58 |
| rc101-5 | 47.40% | 16 | 7.70 | 4.02 | 5.48 | 5.12 | 3.38 | 5.82 | 4.58 |
| rc102-1 | 59.00% | 15 | 1.74 | 1.78 | 1.22 | 0.54 | 0.92 | 1.10 | 1.34 |
| rc102-2 | 57.50% | 15 | 4.28 | 1.94 | 3.44 | 2.76 | 2.12 | 2.86 | 2.6 |
| rc102-3 | 56.00% | 15 | 8.70 | 3.24 | 5.06 | 3.32 | 3.38 | 4.14 | 3.24 |
| rc102-4 | 52.00% | 14 | 2.18 | 0.92 | 1.48 | 1.84 | 1.64 | 1.96 | 1.78 |
| rc102-5 | 57.60% | 15 | 3.76 | 2.46 | 2.90 | 2.02 | 1.52 | 2.68 | 2.28 |
| rc104-1 | 76.10% | 13 | 21.10 | 19.70 | 14.40 | 13.82 | 8.64 | 11.10 | 11.4 |
| rc104-2 | 75.60% | 14 | 25.56 | 28.58 | 13.92 | 11.70 | 20.12 | 11.86 | 10.48 |
| rc104-3 | 76.10% | 13 | 20.90 | 16.64 | 10.40 | 8.84 | 7.88 | 7.80 | 9.06 |
| rc104-4 | 72.20% | 12 | 19.60 | 19.28 | 14.08 | 6.36 | 9.80 | 8.74 | 7.38 |
| rc104-5 | 74.40% | 11 | 15.86 | 18.96 | 14.00 | 9.94 | 11.70 | 10.10 | 8.72 |

**Figure 10.5:** Online Vehicle Routings: The Algorithms on Class 4.

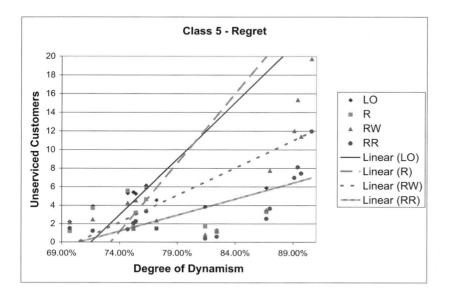

| Problem | DOD | Vehicles | LO | $\mathcal{C}$ | $\mathcal{CW}$ | $\mathcal{CR}$ | $\mathcal{R}$ | $\mathcal{RW}$ | $\mathcal{RR}$ |
|---------|-----|----------|-----|------|------|------|------|------|------|
| rc101-1 | 75.40% | 18 | 5.26 | 4.52 | 4.82 | 2.96 | 3.20 | 4.54 | 2.28 |
| rc101-2 | 74.70% | 17 | 5.32 | 8.08 | 5.04 | 1.92 | 5.58 | 4.26 | 1.40 |
| rc101-3 | 69.70% | 19 | 2.22 | 2.36 | 2.76 | 2.18 | 1.20 | 2.16 | 1.52 |
| rc101-4 | 71.70% | 17 | 3.90 | 6.14 | 2.06 | 1.56 | 3.72 | 2.48 | 1.24 |
| rc101-5 | 76.30% | 16 | 6.14 | 5.56 | 6.54 | 6.00 | 4.66 | 5.98 | 3.36 |
| rc102-1 | 77.20% | 15 | 4.56 | 2.36 | 2.40 | 1.40 | 1.50 | 2.36 | 1.48 |
| rc102-2 | 86.70% | 16 | 5.88 | 5.38 | 2.54 | 2.54 | 3.30 | 3.52 | 2.56 |
| rc102-3 | 81.40% | 17 | 3.84 | 2.40 | 1.06 | 0.16 | 1.72 | 0.84 | 0.40 |
| rc102-4 | 75.20% | 16 | 5.42 | 2.40 | 1.38 | 1.44 | 1.62 | 1.46 | 2.06 |
| rc102-5 | 82.40% | 17 | 1.32 | 1.38 | 1.58 | 0.76 | 1.22 | 1.18 | 0.62 |
| rc104-1 | 89.40% | 14 | 25.26 | 26.90 | 23.02 | 8.76 | 26.14 | 15.34 | 8.12 |
| rc104-2 | 90.60% | 14 | 25.02 | 28.92 | 31.26 | 17.32 | 35.00 | 19.72 | 11.96 |
| rc104-3 | 89.10% | 15 | 26.30 | 27.30 | 19.50 | 7.88 | 24.54 | 12.02 | 6.98 |
| rc104-4 | 89.70% | 14 | 30.18 | 31.18 | 22.02 | 12.38 | 33.86 | 11.42 | 7.44 |
| rc104-5 | 87.00% | 14 | 22.66 | 31.80 | 18.28 | 10.56 | 28.32 | 7.74 | 3.66 |

**Figure 10.6:** Online Vehicle Routing: Unserviced Customers on Class 5.

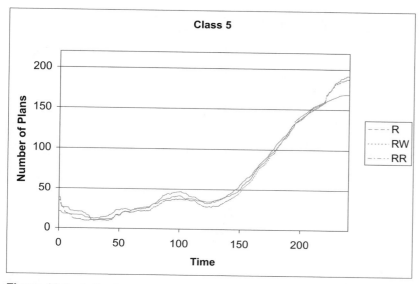

**Figure 10.7:** Online Vehicle Routing: The Number of Plans for the Regret Algorithms.

yet it is critical to obtain high-quality solutions on these instances. The regret algorithm $\mathcal{RW}$ with a waiting strategy is also effective and provides significant benefits over some of the other algorithms, but it is dominated by $\mathcal{RR}$.

## 10.3.2   The Number of Plans over Time

Figure 10.7 shows the number of plans at all times for the regret algorithms on instances of class 5. The results indicate that the regret algorithm maintains a reasonable number of plans over time, although there is time only for five (short) optimizations in between decisions. The exception is at the beginning of the day, where plans may disappear quickly due to the early decisions. The number of plans steeply increases at the end of the day, which is explained easily since no customers arrive in period 3.

## 10.3.3   The Anticipativity Assumption

Figure 10.8 studies the validity of the anticipativity assumption for the regret algorithms on class 5. The results show that, once again, there is significant agreement among the decisions taken in different scenarios. The agreement (that is, the number of matching decisions across scenarios) is around 70 percent most of the time with valleys early in the day and after two hours. These valleys coincide once again with the decrease in the number of plans, which is to be expected, and are

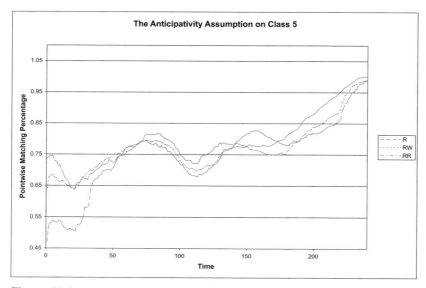

**Figure 10.8:** Online Vehicle Routing: The Anticipativity Assumption for the Regret Algorithms.

lower early in the day. At the end of the day (period 3), no customer arrives, which means that the agreement between scenarios converges to 100 percent.

Figure 10.9 provides some telling results. It depicts the waiting times of the accepted customers in the various algorithms on class 5. Recall that, on class 5, $\mathcal{RR}$ and $\mathcal{RW}$ serve more customers than LO and $\mathcal{R}$, but it is not clear why. The results indicate that the benefits of algorithms $\mathcal{RR}$ and $\mathcal{RW}$ stem from their ability to serve many customers early. By relocating or by waiting in their current location, the vehicles are in a better position to accommodate the customers soon after they request service. This is in fact a desirable property in itself and, once again, an emerging behavior of $\mathcal{RR}$ and $\mathcal{RW}$.

### 10.3.4   Accepting and Rejecting Customers

An important question to study is whether stochastic information may help the algorithm to determine if it is beneficial to reject some customers in the hope of serving more later. Table 10.2 depicts some very preliminary results in that direction. The idea, which is inspired by the work in [25] on online groceries, is to determine whether it is beneficial to accept a request over the whole set of routing plans instead of greedily accepting a customer as soon as it can be inserted in some routing plan. Table 10.2 compares the algorithms $\mathcal{CW}$ and $\mathcal{RW}$ with and without the ability of rejecting customers based on stochastic information. (The variations are superscripted with a star). In gen-

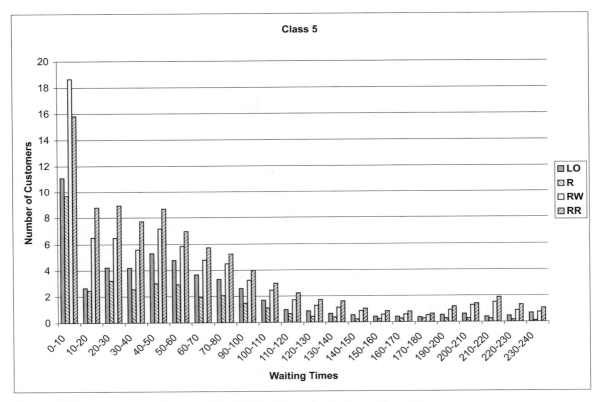

**Figure 10.9:** Online Vehicle Routing: The Waiting Times for the Regret Algorithms.

eral, the results are similar in quality to the results using greedy acceptance. This is due in part to the uniform benefit of servicing customers (which contrasts the problem in [25]), which makes it more attractive to accept any customer now than to wait for future requests. Some improvements are seen for high DODs, but it is not clear whether they are statistically significant.

## 10.4   Notes and Further Reading

It took us five years to derive the results of this chapter. Online vehicle routing with time windows was the first online stochastic problem we studied in order to demonstrate the value of stochastic information. The initial results were reported [12] and were based on the consensus algorithms. Subsequent research led to the regret algorithm (when studying packet scheduling), the waiting strategy (when studying vehicle dispatching) and, more recently, to the relocation strategy when

| Problem | DOD | Vehicles | LO | $\mathcal{CW}$ | $\mathcal{CW}^*$ | $\mathcal{RW}$ | $\mathcal{RW}^*$ |
|---------|------|----------|-------|-------|-------|-------|-------|
| rc104-1 | 76.1% | 13 | 21.10 | 14.40 | 14.98 | 11.10 | 9.38 |
| rc104-2 | 75.6% | 14 | 25.56 | 13.92 | 12.66 | 11.86 | 10.94 |
| rc104-3 | 76.1% | 13 | 20.90 | 10.40 | 12.50 | 7.80 | 8.24 |
| rc104-4 | 72.2% | 12 | 19.60 | 14.08 | 13.64 | 8.74 | 7.76 |
| rc104-5 | 74.4% | 11 | 15.86 | 14.00 | 15.46 | 10.10 | 9.24 |

**Table 10.2:** Results on Class 4 Online Vehicle Routing with Time Windows: Stochastic Accept or Reject.

reconsidering online vehicle routing. Retrospectively, the simplicity of the waiting and relocation strategies is quite remarkable.

Research on modeling uncertainty in vehicle routing and dispatching problems can be divided into two directions: two-stage stochastic programming and online algorithms without stochastic information.

Most research on stochastic vehicle routing minimizes the expected travel distance. In general, a simple recourse function (that is, returning to the depot) is available during execution when feasibility constraints are violated. The recourse function adds some cost to the expectation. See [17, 68] for some early work and [43] for an overview of the various models and approaches. More recently, [115] studied the standard stochastic demand problem and adds the ability to preemptively return to the depot to unload capacity if the expectation makes it a better choice than a likely forced return later. Reference [41] approaches the problem with stochastic demand and stochastic customers and solves the expectation optimally using linear programming techniques. Reference [42] uses the same model but uses tabu search to solve larger instances of the problem. The recourse function in these problems is to return to the depot for unloading whenever the capacity constraint is violated. Reference [83] treats the problem as space partitioning. Each vehicle is assigned regions of space and the goal is to create regions such that the probability of failing to service all customers in the regions is below some threshold. Reference [98] formulates the problem as a stochastic dynamic program and uses reinforcement learning to approximate good solutions.

Most research in online vehicle routing ignores stochastic information. Reference [86] provides an excellent survey of online vehicle routing. The most relevant work is probably the parallel tabu-search algorithm of [40]. This algorithm is based on an adaptive memory that stores potential solutions. As new customer requests arrive, the algorithm uses those solutions that allow service of the customer. The algorithm may violate time windows but there is a penalty incorporated into the objective function for late arrivals. As a consequence, feasibility is greatly simplified, but the objective function is more complex. The algorithm does not exploit stochastic information and uses travel time to choose between different solutions. Thanks to the use of the adaptive memory and the maintenance of multiple solutions, the algorithm significantly outperforms the greedy approach on moderately dynamic problems. It motivated the design and implementation of the LO approach used here, which can be seen as a generalization and abstraction of this algorithm by isolating the component for searching solutions and the ranking function. See also [39], which applies similar ideas to a dynamic pickup-and-delivery problem and suggests adding historical or stochastic information as an avenue of future research. The other most relevant work is the excellent paper [70], which defines the notion of "degree of dynamism", that is, percentage of dynamic customers, and the notion of "effective degree of dynamism," which captures the lateness of the dynamic customers, as well as its generalization to time windows (see also [69]). The experimental setting here is precisely based on these critical concepts. The paper also studies a dynamic problem where service time is an independent random variable. It compares several heuristics: first come/first serve, stochastic queue median, nearest neighbor, and partitioning. The nearest-neighbor heuristic, which is essentially the greedy heuristic used here, performs the best. Observe that the paper also indicates that little work has been done on incorporating stochastic information in a dynamic setting. Reference [67] is another interesting paper that addresses a large (1,600 vehicles) real-world problem. The application is concerned with a large automobile club that provides service for club customers whose vehicles break down. The goal is to minimize a cost function that is a linear combination of operational costs (service, driving, overtime, and contractors) and lateness costs with respect to soft time windows. The authors study how to optimize the objective functions given the known information (the approach thus can be viewed as an LO approach with a single plan). They propose a mixed-integer-programming (MIP) approach based on column generation to find optimal or near-optimal solutions. Note the authors mention that the instances must be sufficiently well behaved for a MIP approach to be successful, which is not the case for the Solomon benchmarks.

References [18, 19] report some of the earliest work in mathematically studying some models of dynamic vehicle routing. They look at a model with Poisson arrival rates for customers that are uniformly distributed in the plane. The goal is to find a policy that minimizes the total system time of an infinite horizon (where system time is, for each customer, the total time spent servicing a customer and the waiting time of a customer). The policies developed here (and a finite time model version of this) are assessed in [70]. Reference [109] uses the same model as [18, 19] and applies similar techniques and analysis to the pickup and delivery problem. Reference [44] presents an algorithm for ambulance relocation, which consists of two subproblems. The first subproblem is to send an ambulance when a call comes in, and the second problem is to relocate ambulances such that areas of the region are covered by at least two ambulances. The objective function is to minimize the region that is not double-covered and the cost of relocation to double-cover more of the region. Referenxe [54] explores the benefits of allowing vehicles to change their destinations (diversion) when traveling. Reference [93] uses integer-programming techniques to optimize the current state of information, while reference [24] decomposes every entity of the problem (i.e., vehicles, customers) into interactive agents working together to satisfy the current state of information. [116] is another agent-driven approach used to react to traffic jams. There has also been some work in proving competitive-ratio bounds for various versions of the VRP. These include [2] for finding an optimal 2-competitive algorithm for the online single-vehicle dial-a-ride problem that minimizes the completion time. Other work on this problem has been performed by [51].

As mentioned earlier, very little work uses stochastic information for dynamic vehicle routing. A notable exception is [99], which considers stochastic demands and only one vehicle. All customers are known and there are no time windows. They apply a rollout algorithm to approximate good solutions every time new information becomes available. More recently, distribution information has been used increasingly often. For example, [25] has considered vehicle routing in the context of Web grocery delivery problems. Potential customers are given choices of potential delivery times. The choices that are presented are limited to those whose expected profit is sufficiently high. This work is extended in [26] to add incentives for customers to choose less popular time slots.

It is worth emphasizing that, in our algorithms, the vehicles can wait for relocate anywhere and at any time during the algorithm execution. This contrasts with earlier approaches (e.g., [71, 111]) where waiting and relocation points are defined a priori using knowledge of the distribution, clustering of the customers, and heuristics. Moreover, a decision to wait or relocate solely relies on the scenario solutions, not on any prior knowledge of the problem or the distribution. As a result, the strategies should naturally transfer to a variety of applications.

# IV LEARNING AND HISTORICAL SAMPLING

# 11 Learning Distributions

*The future is already here — it's just unevenly distributed.*
— William Gibson

*The economists are generally right in their predictions, but generally a good deal out in their dates.*
— Sidney Webb

The online stochastic optimization algorithms presented so far assume the existence of a distribution that can be sampled to generate scenarios of the future. This chapter revisits this assumption for applications in which this assumption does not hold. It shows how to adapt machine-learning techniques to learn the distribution on the fly during the execution of the online algorithms.

This chapter focuses on continuous online optimization and distributions specified as Markov models. Section 11.1 reviews how to enhance the stochastic algorithms with a learning component. Section 11.2 considers the case where the distribution is a hidden Markov model and shows how to use belief states to sample effectively. Then section 11.3 studies the case where the transition probabilities of the hidden Markov model are not available and shows how to use the Baum-Welch algorithm to learn the model online.

## 11.1 The Learning Framework

This chapter focuses on continuous online optimization, such as packet scheduling, in which the online algorithm runs continuously or at least for long periods of time. In these applications, earlier subsequences of inputs reveal information on the distribution and may be used to infer the state of the distribution or to train a partially specified model.

For simplicity, this chapter assumes that the input distribution for the online stochastic problems is a (possibly under-specified) Markov model. The Markov model emits the input for the online algorithms at each time step. However, since this chapter takes an inside look at sampling, it is more intuitive to refer to the emitted symbols as outputs here. A Markov model can be specified by a triple $\mathcal{M} = (S, O, t)$, where $S$ is a set of states, $O$ is a set of outputs, and $p : S \times S \times O \rightarrow [0, 1]$ is a transition probability function, that is, $p(s_t, s_{t+1}, o_{t+1})$ represents the probability of moving from state $s_t$ to state $s_{t+1}$ while emitting $o_{t+1}$. Figure 11.1 depicts a Markov model with three states ($S = \{1, 2, 3\}$), two possible outputs ($O = \{a, b\}$), and whose transition probabilities from state 2 are specified as

t(2,2,a) = 0.95
t(2,2,b) = 0.10
t(2,3,a) = 0.20
t(2,3,b) = 0.20

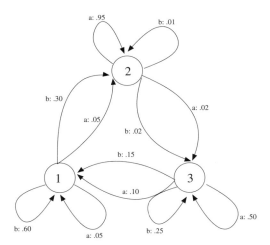

**Figure 11.1:** A Markov Model.

An implementation of the sampling procedure under these assumptions is shown in figure 11.2. The implementation has access to the state $\mathcal{I}_t$ of the distribution at time $t$ since $\mathcal{I}$ is fully observable. To sample the time frame $[b, e]$, the procedure retrieves the state $\mathcal{I}_b$ of the distribution and performs a random walk in the Markov model. At each time $t$, it generates a transition $(s_{t+1}, o_{t+1})$ using the transition probabilities (line 3) and continues from state $s_{t+1}$ (line 5). The output of the sampling procedure is a sequence $\langle o_b, \ldots, o_e \rangle$ that can be used to build a scenario in the online algorithms.

The sampling procedure makes two assumptions. First, it assumes that the distribution $\mathcal{I}$ is fully observable, that is, at each time $t$, the state of $\mathcal{I}$ is known to the sampling procedure. This is a strong assumption, which may be unrealistic in practice. Second, the sampling procedure assumes that the transition probabilities are known and do not vary over time. The rest of this chapter shows how machine-learning techniques, that is, belief states and the Baum-Welch algorithm, address both limitations for continuous online optimization. This requires the online stochastic algorithms to include a learning phase as well. For illustration purposes, figure 11.3 depicts the generalization of the generic online algorithm for online scheduling where the learning phase takes place in line 3 before the sampling in the decision-making process in line 4. The learning receives as input the requests in earlier time steps, which, of course, were generated from the distribution.

## 11.2   Hidden Markov Models

In many applications, the state of a distribution at time $t$ is not observable. In other words, although the distribution is specified by a Markov model, its state at a time $t$ is not known to the

SAMPLE$(b, e)$
1  **return** RANDOMWALK$(I_b, b, e)$;

RANDOMWALK$(s, b, e)$
1  $O \leftarrow \langle \rangle$;
2  **for** $t \in b..e$ **do**
3     SELECT$(s_t, o_t)$ WITH PROBABILITY $p(s, s_t, o_t)$;
4     $O \leftarrow O : o_t$;
5     $s \leftarrow s_t$;
6  **return** $O$;

**Figure 11.2:** The Sampling Procedure for Fully Observable Markov Models.

ONLINE ALGORITHM $\mathcal{A}(\langle R_1, \ldots, R_h \rangle)$
1  $S_0 \leftarrow \langle \rangle$;
2  **for** $t \in H$ **do**
3     LEARN$(\langle R_1, \ldots, R_t \rangle)$;
4     $s_t \leftarrow$ CHOOSEREQUEST$(S_{t-1}, \langle R_1, \ldots, R_t \rangle)$;
5     $S_t \leftarrow S_{t-1} : s_t$;
6  **return** $S_h$;

**Figure 11.3:** The Generic Online Algorithm with a Learning Phase.

online algorithms. The resulting distributions are often called partially observable or hidden Markov models. The sampling procedure shown earlier cannot be used, since the state $\mathcal{I}_b$ is not available.

The online algorithm, however, has access to the sequence of outputs from the distribution that can be used to infer information about the states. More precisely, the algorithm may maintain a belief $B_t(s)$ for each state $s \in S$ representing the probability that the distribution $\mathcal{I}$ be in state $s$ at time $t$. When $S = \{s_1, \ldots, s_n\}$, the vector $\langle B_1(s_1), \ldots, B_n(s_n) \rangle$ is typically called a belief state. The implementation of procedure LEARN and SAMPLE is depicted in figure 11.4. The learning procedure uses the belief state and the output at time $t$ to infer the belief state at time $t + 1$ (lines 1 and 2). The belief $B_{t+1}(s)$ of a state $s$ at time $t + 1$ is given by the probability

$$pr(s | o_t, B_t),$$

LEARN-HMM($o_t$)
1  **for** $s \in S$ **do**
2      $B_{t+1}(s) \leftarrow \Pr(s|o_t, B_t)$;

SAMPLE-HMM($b, e$)
1    SELECT $s \in S$ WITH PROBABILITY $B_b(s)$;
2    **return** RANDOMWALK($s, b, e$);

**Figure 11.4:** The Learning and Sampling Algorithms for Hidden Markov Models.

which, by Bayes' formula, reduces to

$$\frac{\sum_{s' \in S} B_{t-1}(s') \, p(s', s, o_t)}{\sum_{s' \in S} \sum_{s'' \in S} B_{t-1}(s') \, p(s', s'', o_t)}.$$

Once the belief state is available, the sampling procedure randomly generates a state $s$ according to the probabilities in the belief state (line 1) and proceeds with a traditional random walk (line 2) from that state. Different samples may thus use different starting states for the random walk.

Figure 11.5 depicts the experimental results of the online stochastic algorithms for HMMs. It compares the stochastic algorithms when the distribution is a (fully observable) Markov model and a hidden Markov model. The results show that sampling from the belief states degrades the quality of the algorithms only marginally. For algorithms $\mathcal{R}$ and $\mathcal{E}$, the loss in quality is negligible, while the loss for algorithm $\mathcal{C}$ is slighly more significant. Moreover, the benefits of the online stochastic algorithms over the oblivious algorithms remain significant in all cases, indicating that specifying the distribution as a hidden Markov model does not reduce the benefits of the online stochastic approach.

## 11.3  Learning Hidden Markov Models

Consider now the case where the distribution $\mathcal{I}$ is a hidden Markov model whose transition probabilities are not known precisely or may evolve over time. The online algorithms must infer from the output sequence not only the belief states, but also the transition probabilities.

Assume first that the Markov model is fully observable but the transition probabilities are unknown.[1]  Given a state sequence and an output sequence $o_n$, procedure COUNT in figure 11.6

---

[1] This is entirely unrealistic, but this assumption helps convey the intuition.

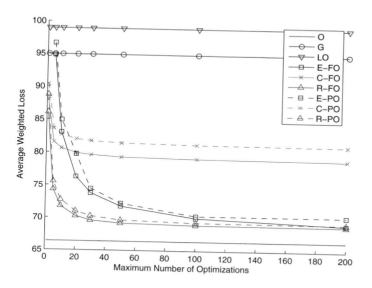

**Figure 11.5:** Online Packet Scheduling with Hidden Markov Models.

estimates the transition probability $p(s, s', o')$ from $s_n$ and $o_n$. This estimation can be performed on the fly at each time $t$ as the online algorithm proceeds as shown in figure 11.7. Figure 11.8 shows that this estimation is really effective for online packet scheduling. It compares the online stochastic algorithms with the (fully observable) Markov model and with the Markov models in which the transition probabilities are estimated as described in figure 11.7. As can be seen from the figure, there is essentially no difference in quality between the exact and estimated Markov models.

Unfortunately, for hidden Markov models the states $\mathcal{I}_{t-1}$ and $\mathcal{I}_t$ are not observable and the implementation in figure 11.7 cannot be used. The Baum-Welch algorithm [4] overcomes this limitation by estimating the transition probabilities iteratively from the belief states and initial guesses. Following [29], this section derives the Baum-Welch algorithm systematically, assuming first that the initial state $s_1$ is known.

One of the main ideas of the Baum-Welch algorithm is to estimate the transition probabilities by summing over all state sequence $s_{n+1}$. The estimations then become

$$
\begin{aligned}
\hat{C}(s, s', o') &= \sum_{s_{n+1}} \Pr(s_{n+1} | o_n) \, \text{COUNT}(s, s', o', s_{n+1}, o_n) \\
&= \frac{1}{\Pr(o_n)} \sum_{s_{n+1}} \Pr(s_{n+1}, o_n) \, \text{COUNT}(s, s', o', s_{n+1}, o_n).
\end{aligned}
$$

COUNT$(s, s', o', \boldsymbol{s_{n+1}}, \boldsymbol{o_n})$
1    **for** $t \in 1..n$ **do**
2       $C(s_t, s_{t+1}, o_t) \leftarrow C(s_t, s_{t+1}, o_t) + 1$;
3    **for** $s \in S$ **do**
4       $T \leftarrow \sum_{s' \in S, o' \in O} C(s, s', o')$;
5       **for** $s' \in S, o' \in O$ **do**
6          $p(s, s', o') \leftarrow \frac{C(s, s', o')}{T}$
7    **return** $p(s, s', o')$;

**Figure 11.6:** Estimating the Transition Probabilities of a Markov Model.

LEARN-MM-TRANSITIONS$(\boldsymbol{o_t})$
1    $C(\mathcal{I}_{t-1}, \mathcal{I}_t, o_t) \leftarrow C(\mathcal{I}_{t-1}, \mathcal{I}_t, o_t) + 1$;
2    $T \leftarrow \sum_{s \in S, o \in O} C(\mathcal{I}_{t-1}, s, \omega)$;
3    **for** $s \in S, \omega \in \Omega$ **do**
4       $p(\mathcal{I}_{t-1}, s, o) \leftarrow \frac{C(\mathcal{I}_{t-1}, s, o)}{T}$;

**Figure 11.7:** Estimating the Transition Probabilities of a Markov Model.

The transition probabilities can then be estimated by

$$p(s, s', o') = \frac{\hat{C}(s, s', o')}{\sum_{s'' \in S} \sum_{o'' \in O} \hat{C}(s, s'', o'')}. \tag{11.3.1}$$

Since the term $\frac{1}{\Pr(\boldsymbol{o_n})}$ cancels, the rest of this section abuses notation and uses

$$\hat{C}(s, s', o') = \sum_{\boldsymbol{s_{n+1}}} \Pr(\boldsymbol{s_{n+1}}, \boldsymbol{o_n}) \text{ COUNT}(s, s', o', \boldsymbol{s_{n+1}}, \boldsymbol{o_n})$$

for simplicity. Obviously, it is not practical to enumerate all state sequences. Fortunately, $\hat{C}(s, s', o')$ can be rewritten as

$$\hat{C}(s, s', o') = \sum_{t=1}^{n} \Pr(s_t = s, s_{t+1} = s', o_t = o', \boldsymbol{o_n}),$$

which sums the probabilities of performing the transition $(s, s', o')$ at each time $t$. These estimations can be rewritten into a form amenable to efficient implementation by splitting $\boldsymbol{o_n}$ into $o_{1..t}$ and

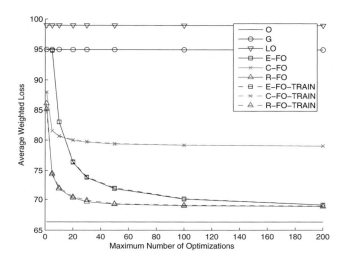

**Figure 11.8:** Online Packet Scheduling: Training of Transition Probabilities in Markov Models.

$o_{t+1..n}$. We have

$$
\begin{aligned}
\hat{C}(s, s', o') &= \sum_{t=1}^{n} \Pr(s_t = s, s_{t+1} = s', o_t = o', \boldsymbol{o_n}) \\
&= \sum_{t=1}^{n} \Pr(o_{1..t-1}, s_t = s, s_{t+1} = s', o_t = o', o_{t+1..n}) \\
&= \sum_{t=1}^{n} \Pr(o_{1..t-1}, s_t = s) \Pr(s_{t+1} = s', o_t = o'|o_{1..t-1}, s_t = s) \\
&\quad \Pr(o_{t+1..n}|o_{1..t-1}, s_t = s, s_{t+1} = s') \\
&= \sum_{t=1}^{n} \Pr(o_{1..t-1}, s_t = s) \Pr(s_{t+1} = s', o_t = o'|s_t = s) \Pr(o_{t+1..n}|s_{t+1} = s') \\
&= \sum_{t=1}^{n} \Pr(o_{1..t-1}, s_t = s) \, p(s, s', o') \Pr(o_{t+1..n}|s_{t+1} = s').
\end{aligned}
$$

We now show how to compute $\Pr(o_{1..t-1}, s_t = s)$ and $\Pr(o_{t+1..n}|s_{t+1} = s')$ with the so-called forward and backward algorithms.

**The Forward Algorithm**   Denote by

$$\alpha(s,t) = \Pr(s_t = s, \boldsymbol{o_{t-1}})$$

the probability of ending up in state $s$ at time $t$ from state $s_1$ given the sequence $\boldsymbol{o_{t-1}}$. The forward algorithm computes $\alpha(s,t)$ by induction on $t$. The base case is

$$\alpha(s,1) = \begin{cases} 1 & \text{if } s = s_1 \\ 0 & \text{otherwise.} \end{cases}$$

The recursive case is derived as follows:

$$
\begin{aligned}
\alpha(s,t+1) &= \Pr(\boldsymbol{o_t}, s_{t+1} = s) \\
&= \sum_{s' \in S} \Pr(s_t = s', s_{t+1} = s, \boldsymbol{o_t}) \\
&= \sum_{s' \in S} \Pr(s_t = s', \boldsymbol{o_{t-1}}) \Pr(s_{t+1} = s, o_t | \boldsymbol{o_{t-1}}, s_t = s') \\
&= \sum_{s' \in S} \Pr(s_t = s', \boldsymbol{o_{t-1}}) \Pr(s_{t+1} = s, o_t | s_t = s') \\
&= \sum_{s' \in S} \alpha(s',t)\, p(s', s, o_t).
\end{aligned}
$$

As a result, $\alpha(s,t)$ can be computed efficiently using dynamic programming.

**The Backward Algorithm**   The backward algorithm is the dual of the forward algorithm. Denote by

$$\beta(s,t) = \Pr(o_{t..n} | s_t = s)$$

the probability of emitting $o_{t..n}$ when starting from state $s$ at time $t$. The base case is defined as

$$\forall s \in S : \beta(s, n+1) = 1.$$

The recursive case can be derived as in the forward algorithm to obtain

$$\beta(s, t-1) = \sum_{s' \in S} p(s, s', o_{t-1})\beta(s', t),$$

which can also be computed efficiently by dynamic programming.

**The Baum-Welch Algorithm**   Putting all the results together, it follows that

$$\hat{C}(s, s', o') = \sum_{t=1}^{n} \alpha(s,t)\, p(s, s', o')\, \beta(s', t+1).$$

LEARN-HMM-T$(o_t)$
1   **for** $s \in S$ **do**
2      $B_{t+1}(s) \leftarrow \Pr(s|o_t, B_t)$;
3   **repeat**
4      $po \leftarrow \Pr(o_{t-\lambda+1..t})$;
5      **for** $s \in S$ **do**
6        $\alpha(s, t - \lambda + 1) = B_{t-\lambda+1}(s)$;
7        **for** $i \leftarrow t - \lambda + 1$ TO $t$ **do**
8          $\alpha(s, i + 1) \leftarrow \sum_{s' \in S} \alpha(s', i) \, p(s', s, o_i))$;
9        $\beta(s, t + 1) = B_{t+1}(s)$;
10      **for** $i \leftarrow t + 1$ DOWNTO $t - \lambda + 3$ **do**
11        $\beta(s, i - 1) \leftarrow \sum_{s' \in S} p(s, s', o_{i-1}) \, \beta(s', i)$;
12      **for** $s \in S, s' \in S, o' \in O$ **do**
13        $\widehat{C}(s, s', o) \leftarrow \sum_{i=t-\lambda+1}^{t} \alpha(s, i) \, p(s, s', o') \, \beta(s', i + 1)$;
14      **for** $s \in S, s' \in S, o' \in O$ **do**
15        $p(s, s', o') \leftarrow \widehat{C}(s, s', o') / \sum_{s'' \in S, o'' \in O} \widehat{C}(s, s'', w'')$;
16  **until** $|\Pr(o_{t-\lambda+1..t}) - po| \leq \epsilon$;

SAMPLE-HMM-T$(b, e)$
1  SELECT $s \in S$ WITH PROBABILITY $B_b(s)$;
2  **return** RANDOMWALK$(s, b, e)$;

**Figure 11.9:** The Learning and Sampling Algorithms for Trained Hidden Markov Models.

These estimations, in conjunction with formula (11.3.1), give a way to re-estimate the transition probability $p(s, s', o')$ from an initial guess. The Baum-Welch algorithm iterates such steps until the transition probabilities do not evolve significantly or, alternatively, until the probability $\Pr(o_n)$ does not change much. This last probability is easy to estimate using the forward algorithm, and is given by

$$\sum_{s \in S} \alpha(s, n + 1).$$

**Training the Hidden Markov Models Online**  We now show how to train the hidden Markov models in an online setting. The main idea is to train the hidden Markov models using the Baum-Welch algorithm on the recent output sequence $o_{t-\lambda+1..t}$, where $\lambda$ is a parameter denoting the length of the sequence. Moreover, the base case in the forward and backward algorithms is initialized with

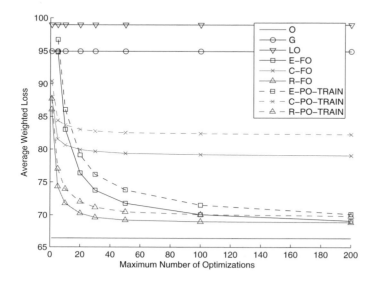

**Figure 11.10:** Online Packet Scheduling: Potential Quality of Online Learning of the Hidden Markov Model.

the belief state of each state $s \in S$ at time $t - \lambda + 1$ and $t + 1$, that is,

$$\alpha(s, t - \lambda + 1) = B_{t-\lambda+1}(s)$$
$$\beta(s, t + 1) = B_{t+1}(s).$$

The learning and sampling algorithms are depicted in figure 11.9. The sampling procedure is similar to the case of hidden Markov models. The learning procedure starts by computing the belief state at time $t + 1$ (lines 1 and 2). It then applies the forward algorithm (lines 6 through 8) and the backward algorithm (lines 9 through 11) for each state. The counts $\hat{C}$ are computed in lines 12 and 13 and the transition probabilities are re-estimated in lines 14 and 15. These steps are repeated until the probability $\Pr(o_{t-\lambda+1..t})$ of the sequence does not change substantially.

**Experimental Results on Packet Scheduling**   Figure 11.10 depicts the experimental results on packet scheduling when the HMM is learned online. To obtain high-quality results, the parameter $\lambda$, that is, the length of the subsequences used for training, was set to 560. The figure shows that the loss in quality is small for algorithms $\mathcal{R}$ and $\mathcal{E}$ and a bit more significant for $\mathcal{C}$. The loss is larger than for packet scheduling with known HMMs, but, once again, the difference is not substantial. The algorithms still bring significant benefits compared to the oblivious algorithms.

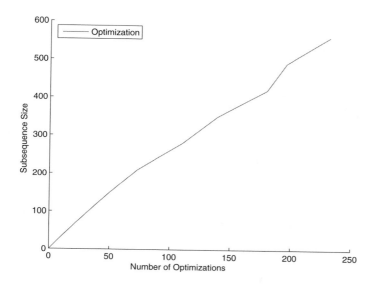

**Figure 11.11:** Learning Time Versus Optimization Time on Packet Scheduling.

However, the results in figure 11.10 do not include the learning time and hence do not reflect an actual execution of the algorithm: they give information only on the potential accuracy of the online HMM learning. Figure 11.11 depicts the computation times required to learn the HMMs as a function of $\lambda$, that is, the size of the subsequence of outputs used to learn the HMM. In the figure, the computation times for learning are given in terms of the number of optimizations in order to characterize the impact of learning on the quality of the algorithms under time constraints. The figure indicates that learning with subsequences of size $\lambda = 500$ is roughly equivalent to 200 optimizations, which is substantial for this application. Indeed figure 11.12 depicts the behavior of the online stochastic algorithms when the learning time is included. It indicates that the algorithms need the time equivalent of 250 optimizations before they become effective, which makes it ineffective for tight time constraints.

The situation may be improved by using shorter training subsequences, albeit at a loss in quality. Figure 11.13 shows the same results when $\lambda = 140$. The results indicate that algorithm $\mathcal{R}$ now requires time for about 50 optimizations in order to improve the oblivious algorithms. It also indicates that the overall quality of the stochastic algorithms now decreases more significantly, although they still provide improvements over the oblivious algorithms.

A more effective approach is to amortize the cost of learning over time and to apply the online

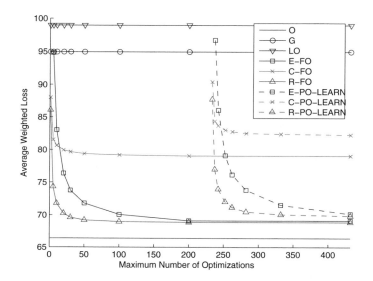

**Figure 11.12:** Online Packet Scheduling: Online Learning of the Hidden Markov Model ($\lambda = 560$).

learning algorithm only periodically instead of at each time step. The results are shown in figure 11.14 and they are promising. The loss in quality is only slightly larger than in the case of known HMMs, and the algorithms perform well under tight time constraints. Once again, the loss in quality in algorithms $\mathcal{R}$ and $\mathcal{E}$ is small, while it is somewhat larger in the case of algorithm $\mathcal{C}$. The online stochastic algorithms also provide significant improvements over the oblivious algorithms.

## 11.4   Notes and Further Reading

A preliminary version of this chapter appears in [14]. The concept of belief states is fundamental in research on partially observable Markov decision processes. See [55] for an excellent starting point on POMDPs. The derivation of the Baum-Welch algorithm was inspired by the beautiful presentation in [29] which also discusses the limitations of the algorithm including the importance of the training sequence and the fact that the result is only a local minimum in general.

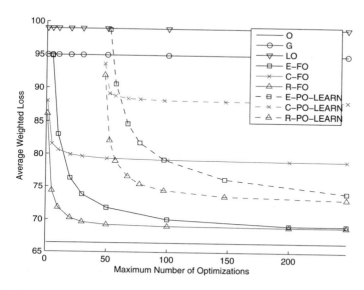

**Figure 11.13:** Online Packet Scheduling: Online Learning of the Hidden Markov Model ($\lambda = 140$).

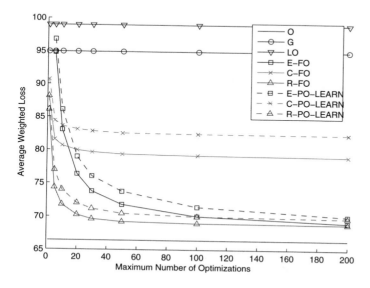

**Figure 11.14:** Online Packet Scheduling: Periodic Online Learning of the Hidden Markov Model.

# 12 Historical Sampling

*Telling the future by looking at the past assumes that conditions remain constant. This is like driving a car by looking in the rearview mirror.*
— Herb Brody

*History may not repeat itself, but it does rhyme a lot.*
— Mark Twain

Chapter 11 showed how to learn distribution during the execution of the online algorithms. The learning algorithms required some knowledge about the distributions, such as the structure of a hidden Markov model. This chapter relaxes these assumptions further and studies applications for which only historic data is available. In continuous optimization, such information is readily available, since earlier input sequences represent historical data. These subsequences are particularly useful when the underlying (unknown) distribution is ergodic.[1] In periodic optimization, it is often the case that corporations have access to a wealth of information about their customers, how they place orders and when, and how this may vary over time. It is thus reasonable to assume the existence of past instances that reasonably resemble the online problem to solve.

This chapter studies how to exploit historical data in the implementation of the sampling procedure whenever a predictive model is not available. The focus is on continuous optimization but the adaptation of the algorithm to periodic optimization is discussed whenever relevant. Moreover, the experimental results are given for both continuous and periodic optimization. Section 12.1 presents historical averaging and discusses its limitation. The concept of historical sampling is then presented and experimental results on packet scheduling and vehicle routing are given.

## 12.1 Historical Averaging

Historical averaging consists of computing the emission frequencies for each output.[2] The idea is to take the sequence of the last $\lambda$ outputs and to estimate the probabilities of the possible outputs using their frequencies. The implementation of historical averaging is shown in figure 12.1 and it uses the separation between learning and sampling from chapter 11. The learning procedure estimates the probabilities in lines 5 and 6, after having counted the occurrences of each output in lines 1 through 4. The sampling procedure generates a sequence of outputs by choosing each output according to the estimated probability distribution.

---

[1]Informally speaking, an ergodic distribution is a random process whose behavior does not depend on the initial conditions and whose statistical properties do not vary with time [30].

[2]Once again, this chapter takes an insider view of sampling and hence the inputs of the scenarios are the outputs of the sampling procedure.

LEARN-HA($o_t$)

1   **for** $o \in O$ **do**
2       $C(o) \leftarrow 0$;
3   **for** $i \leftarrow t - \lambda + 1 \ldots t$ **do**
4       $C(o_i) \leftarrow C(o_i) + 1$;
5   **for** $o \in O$ **do**
6       $p(o) \leftarrow \frac{p(o))}{\lambda}$;

SAMPLE-HA($b, e$)

1   $O \leftarrow \langle \rangle$;
2   **for** $t \in b..e$ **do**
3       $o_t \leftarrow$ SELECT $o \in O$ WITH PROBABILITY $p(o)$;
4       $O \leftarrow O : o_t$;
5   **return** $O$;

**Figure 12.1:** The Implementation of Historical Averaging.

Figure 12.2 depicts the experimental results for historical averaging on packet scheduling for the value $\lambda = 500$, which gives the best results. The results indicate that the online stochastic algorithms with historical averaging provide significant benefits over the oblivious algorithms. However, the loss in quality for algorithms $\mathcal{E}$ and $\mathcal{R}$ is more important than for the learning approaches in chapter 11. The loss for algorithm $\mathcal{C}$ is essentially negligible.

## 12.2   Historical Sampling

Historical averaging captures the frequencies of the outputs but it loses the structure of sequence. Historical sampling recovers this information by using earlier subsequences as samples. Its implementation is shown in figure 12.3. The figure depicts the implementation of the sampling procedure, which now receives an additional input: the sequence of outputs emitted up to time $b$. Its implementation simply generates a random number between 1 and $2b - e$ and returns the subsequence $o_{t..t+e-b}$. Despite its simplicity, historical sampling enjoys some nice benefits:

1. Contrary to historical averaging, historical sampling captures structural information on the sequence.

2. Whenever the underlying (unknown) distribution is a hidden Markov model, historical sampling can be viewed as a random walk in the model starting at a random state. As a result, historical sampling is essentially equivalent to the sampling of the hidden Markov model for

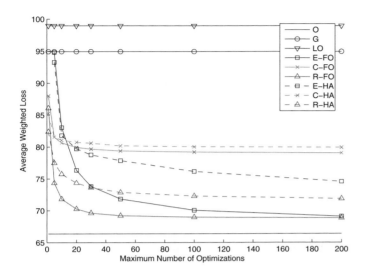

**Figure 12.2:** Experimental Results on Packet Scheduling for Historical Averaging.

SAMPLE-HS($o_{b-1}, b, e$)
1   $t \leftarrow$ RANDOM($[0, 2b - e]$);
2   **return** $o_{t..t+e-b}$;

**Figure 12.3:** The Implementation of Historical Sampling.


which the belief states are "approximated" using the frequencies of the outputs.

3. For periodic optimization, historical sampling consists of selecting a past instance and using the sequence of outputs in this instance in the considered time interval. For instance, in online vehicle routing, historical sampling generates the customer requests in past instances within the time interval.

As a result, historical sampling is a simple technique to exploit historical data in both continuous and periodic optimization.

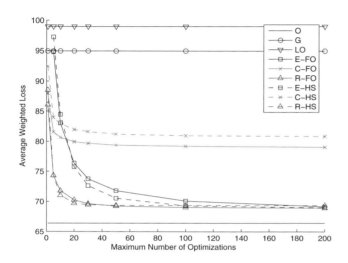

**Figure 12.4:** Experimental Results on Packet Scheduling for Historical Sampling.

### 12.2.1   Experimental Results on Packet Scheduling

Figure 12.4 depicts the experimental results for historical sampling on packet scheduling. The quality of the algorithms is quite remarkable. Algorithms $\mathcal{R}$ and $\mathcal{C}$ suffer no quality loss compared to a (fully observable) Markov model. The consensus algorithm suffers a small loss comparable to those experienced in the learning algorithms. In other words, without knowledge of the distribution and without inducing any time overhead, historical sampling is as effective as the sampling of the actual underlying distribution.

### 12.2.2   Vehicle Routing

Consider now the experimental evaluation of historical sampling on online multiple vehicle routing with time windows, which is a significantly more complicated problem. It is also a periodic optimization problem for which historical data is typically available.

**Experimental Setting**   The experimental results are based on the class-4 problems from chapter 10. These problems are challenging, involve one hundred customers, and exhibit various degrees of dynamism, different distributions of early and late requests, and time windows of different sizes. Hence, they cover a wide spectrum of possibilities and structures.

   Historical sampling was given either 1, 10, 100, or 1,000 historical instances. The sampling

| Problem | C | HS(C)$^1$ | H(C)$^{10}$ | H(C)$^{100}$ | H(C)$^{1,000}$ |
|---|---|---|---|---|---|
| 20-20-60-rc101-1 | 4.16 | 3.50 | 3.28 | 4.28 | 2.36 |
| 20-20-60-rc101-2 | 5.94 | 5.82 | 4.62 | 4.46 | 4.28 |
| 20-20-60-rc101-3 | 3.06 | 3.44 | 2.96 | 2.96 | 2.34 |
| 20-20-60-rc101-4 | 4.30 | 7.22 | 4.30 | 4.98 | 6.00 |
| 20-20-60-rc101-5 | 5.48 | 7.42 | 6.32 | 6.68 | 4.70 |
| 20-20-60-rc102-1 | 1.22 | 0.88 | 1.24 | 0.82 | 1.50 |
| 20-20-60-rc102-2 | 3.44 | 4.14 | 4.28 | 2.70 | 3.92 |
| 20-20-60-rc102-3 | 5.06 | 4.52 | 3.54 | 3.40 | 3.62 |
| 20-20-60-rc102-4 | 1.48 | 2.92 | 1.58 | 1.76 | 1.46 |
| 20-20-60-rc102-5 | 2.90 | 3.64 | 1.48 | 2.28 | 2.24 |
| 20-20-60-rc104-1 | 14.40 | 20.62 | 16.00 | 16.90 | 15.40 |
| 20-20-60-rc104-2 | 13.92 | 15.56 | 15.48 | 15.46 | 15.84 |
| 20-20-60-rc104-3 | 10.40 | 14.86 | 12.34 | 13.00 | 12.76 |
| 20-20-60-rc104-4 | 14.08 | 17.22 | 14.98 | 14.28 | 14.86 |
| 20-20-60-rc104-5 | 14.00 | 16.00 | 13.66 | 16.46 | 14.36 |

**Table 12.1:** Historical Sampling for Online Vehicle Routing (Class 4) for the Consensus Algorithm.

procedure at time $t$ randomly selects a historical instance and returns all requests arriving after $t$ in the instance. Historical sampling loses two kinds of information with respect to the exact distribution. First, it works from a limited pool of instances that may introduce a bias toward certain features. Second, it loses some dependency information. Indeed when the distribution is known, the arrival of a request in a region for a given time period eliminates future requests from that region within the same time period.

**Experimental Results**   Tables 12.1 and 12.2 depict the experimental results for algorithms $\mathcal{C}$ and $\mathcal{R}$ with historical sampling. They compare the number of missed customers when the algorithms use the actual distribution (columns C and R) and historical sampling (columns HS(C) and HS(R)). Each entry in the table is the average of 50 independent runs with 5 optimizations between events. Historical sampling was evaluated for pools of instances of size 1, 10, 100, and 1,000 to assess the importance of the pool size.

The results indicate that historical sampling performs very well on these problems. When the historical data is large, there is no significant difference between using historical sampling or the actual distribution. In general, the quality of the algorithms improves with the pool size, although there are some outliers occasionally. Indeed it may happen that the algorithm receives samples reasonably similar (respectively different) to the current instance, in which case the quality may

| Problem | R | $H(R)^1$ | $H(R)^{10}$ | $H(R)^{100}$ | $H(R)^{1,000}$ |
|---|---|---|---|---|---|
| 20-20-60-rc101-1 | 3.72 | 4.02 | 2.60 | 3.82 | 2.34 |
| 20-20-60-rc101-2 | 4.18 | 5.94 | 4.74 | 4.24 | 4.76 |
| 20-20-60-rc101-3 | 3.46 | 3.12 | 3.28 | 2.42 | 2.92 |
| 20-20-60-rc101-4 | 4.86 | 7.20 | 4.92 | 5.44 | 5.46 |
| 20-20-60-rc101-5 | 5.82 | 7.04 | 5.24 | 6.68 | 4.46 |
| 20-20-60-rc102-1 | 1.10 | 1.34 | 1.22 | 1.02 | 1.20 |
| 20-20-60-rc102-2 | 2.86 | 3.84 | 3.12 | 2.94 | 3.02 |
| 20-20-60-rc102-3 | 4.14 | 4.28 | 3.06 | 3.18 | 2.98 |
| 20-20-60-rc102-4 | 1.96 | 3.04 | 1.80 | 2.08 | 1.52 |
| 20-20-60-rc102-5 | 2.68 | 2.74 | 1.74 | 1.56 | 2.04 |
| 20-20-60-rc104-1 | 11.10 | 12.16 | 10.74 | 11.84 | 10.48 |
| 20-20-60-rc104-2 | 11.86 | 13.24 | 11.22 | 12.66 | 12.60 |
| 20-20-60-rc104-3 | 7.80 | 9.18 | 8.96 | 7.56 | 8.46 |
| 20-20-60-rc104-4 | 8.74 | 13.16 | 11.88 | 10.42 | 9.80 |
| 20-20-60-rc104-5 | 10.10 | 12.56 | 10.18 | 11.16 | 9.88 |

**Table 12.2:** Historical Sampling for Online Vehicle Routing (Class 4) for the Regret Algorithm.

improve (respectively degrade). In particular, algorithms $C$ and $R$ miss fewer customers for 87% and 93% of the instances when the pool size increases from 1 to 10. They miss fewer customers in 40% and 46% of the instances when the pool size is 100.

# V SEQUENTIAL DECISION MAKING

# 13 Markov Chance-Decision Processes

Pascal Van Hentenryck, Luc Mercier, and Eli Upfal

*It is a mistake to look too far ahead. Only one link of the chain of destiny can be handled at a time.*
— Winston Churchill

*I have seen the future, and it is still in the future.*
— James Gleick

Online stochastic scheduling, reservation, and routing represent three classes of applications with similar features. In all three cases, they are specified in terms of an underlying optimization problem, which must be solved online for input sequences drawn from a distribution. This chapter takes a step back: it recognizes that all these application areas, despite the different nature of their decisions and requirements, share the same abstract structure and also are solved by the same abstract algorithms. To formalize this insight, this chapter proposes the concept of the Markov Chance-Decision Process (MCDP) and studies its relationship to the traditional Markov Decision-Chance Process (MDCP). The online anticipatory algorithms can then be reformulated in this abstract setting, highlighting the fundamental computational elements of online stochastic combinatorial optimization: the underlying optimization problem, the anticipatory relaxation, the $\epsilon$-anticipativity assumption, and the clean separation between the optimization and uncertainty models. This chapter also discusses possible directions when the $\epsilon$-anticipativity assumption does not hold, studies its limitations, and reviews opportunities for future research.

## 13.1 Motivation

Markov decision processes (MDPs), and their partially observable variants, are often used to model sequential decision making. They model naturally applications in which the uncertainty depends on the actions, which is typically the case in control and robotics. However, in online stochastic combinatorial optimization, the uncertainty is exogenous. The goal is to maximize the rewards obtained by serving requests and the uncertainty depends only on which requests come and when.

This chapter presents the concept of MCDPs, a variant of MDPs where the uncertainty is exogenous. In MCDPs, at time $t$, the decision process is in a state $s_t$, observes the realization of input $i_t$, and selects the next state $s_{t+1}$. This contrasts with traditional MDPs, called MDCPs here, where the decision process at time $t$ selects an action $a_t$ and observes its uncertain outcome $s_{t+1}$. This chapter shows that MCDPs and MDCPs have the same expressive power.

The benefits of MCDPs for online stochastic combinatorial optimization are computational: they can be tackled online using an abstract version of the anticipatory algorithms presented in this book. Indeed, because the uncertainty is exogenous, MCDPs naturally allow for the anticipatory relaxation that removes the interleaving of decisions and observations and is expressed in terms of deterministic optimization problems. The anticipatory algorithms can thus exploit the anticipatory relaxation on scenarios of the future in order to make decisions online. Moreover, under $\epsilon$-anticipativity assumption, the expected loss of online anticipatory algorithms compared to the offline, a posteriori optimal solution is shown theoretically to be small in this abstract setting. The core of this chapter introduces the concept of MCDPs and studies their expressiveness. It then introduces the online anticipatory algorithms, studies their theoretical performance, and discusses the $\epsilon$-anticipativity assumption. The results focus on MDPs, although many of the results can be generalized to their partially observable variants.

## 13.2  Decision-Chance versus Chance-Decision

This section specifies the concepts of MDCPs and MCDPs. For simplicity, the probability distributions in this chapter are on finites sets. For a set $X$, $\mathrm{prob}(X)$ denotes the set of probability distributions on $(X, 2^X)$. When $x \in X$ and $\pi \in \mathrm{prob}(X)$, we often write $\pi(x)$ instead of $\pi(\{x\})$.

> **Definition 13.1 (Markov Decision-Chance Process)** A Markov Decision-Chance Process is a tuple $(S, s_0, A, T, R)$, where
>
> 1. $S$ is the finite set of states;
> 2. $s_0 \in S$ is the initial state;
> 3. $A$ is the finite set of actions;
> 4. $T : S \times A \to \mathrm{prob}(S)$ is the transition map; and
> 5. $R : S \times A \times S \to \mathbb{R}$ is the reward map.

Running an MDCP consists of choosing an action $a_t \in A$ at time $t$, causing a transition from state $s_t$ to a state whose distribution is $T(s_t, a_t)$. An MDCP *run* is a sequence (finite or not) alternating states and actions, denoted by

$$\left\langle s_0 \xrightarrow{a_0} s_1 \xrightarrow{a_1} \cdots \right\rangle.$$

A run is said *valid* if $\forall t : T(s_t, a_t)(s_{t+1}) > 0$. Let $\mathcal{S} = \langle a_0, \ldots, a_{h-1} \rangle$ be a finite sequence of actions. The probability of a valid run

$$\left\langle s_0 \xrightarrow{a_0} s_1 \xrightarrow{a_1} \cdots \xrightarrow{a_{h-1}} s_h \right\rangle$$

is given by $\prod_{0 \leq t < h} T(s_t, a_t)(s_{t+1})$.

> **Definition 13.2 (Policies for MDCPs)** A policy $\pi$ for an MDCP $(S, A, T, R)$ is a map $\pi : S \to \mathrm{prob}\, A$. The set of policies of an MDCP $\mathcal{M}_\mathcal{A}$ is denoted $\mathrm{pol}(\mathcal{M}_\mathcal{A})$.

Consider a policy $\pi$ for an MDCP $\mathcal{M}_\mathcal{A}$. Let $\mathcal{R}$ be the run

$$\mathcal{R} = \left\langle s_0 \xrightarrow{a_0} s_1 \xrightarrow{a_1} \cdots \xrightarrow{a_{h-1}} s_h \right\rangle.$$

The probability of $\mathcal{R}$ under $(\mathcal{M}_\mathcal{A}, \pi)$ is defined as

$$\prod_{0 \le t < h} \pi(s_t)(a_t) \cdot T(s_t, a_t)(s_{t+1}).$$

**Definition 13.3 (Markov Chance-Decision Process)** A Markov Chance-Decision Process is a tuple $(S, s_0, \Xi, T, R)$, where

1. $S$ is the finite set of states;
2. $s_0 \in S$ is the initial state;
3. $\Xi = (\xi_t)_{t \ge 0}$ is the input process. It is a time-homogeneous Markov chain on a state set $I$ (the input states) specified by an initial distribution $\mu \in \mathrm{prob}(I)$ and a transition matrix $Q$;
4. $T : S \times I \to 2^S$ is the transition map that satisfies $\forall (s, i) \in S \times I : T(s, i) \ne \emptyset$; and
5. $R : S \times I \times S \to \mathbb{R}$ is the reward map.

The semantics of an MCDP is as follows. The Markov chain $(\xi_t)_{t \ge 0}$ is running independently from the behavior of the decision maker. At each time $t$, the decision maker is in state $s_t$, observes the input $i_t$ that is a realization of $\xi_t$, and chooses a new state $s_{t+1} \in T(s_t, i_t)$. An MCDP *run* is a sequence alternating states and observations, denoted by

$$\left\langle s_0 \xrightarrow{i_0} s_1 \xrightarrow{i_1} \cdots \right\rangle,$$

and is *valid* if $\forall t : s_{t+1} \in T(s_t, i_t)$.

**Definition 13.4 (Policies for MCDPs)** A policy $\pi$ for an MCDP $(S, s_0, \Xi, T, R)$ is a map from $S \times I$ to the set of probability distributions over $S$, such that

$$\forall (s, i) \in S \times I, \quad \pi(s, i)(T(s, i)) = 1.$$

The set of policies of an MCDP $\mathcal{M}_\mathcal{A}$ is denoted $\mathrm{pol}(\mathcal{M}_\mathcal{A})$.

Let $\mathcal{M}_\mathcal{A}$ be an MCDP, $\pi$ be a policy for $\mathcal{M}_\mathcal{A}$, and $\mathcal{R}$ be a run of length $h$. The probability of $\mathcal{R}$ under $(\mathcal{M}_\mathcal{A}, \pi)$ is given by

$$\mu(i_0) \cdot \prod_{0 \le t < h} \pi(s_t, i_t)(s_{t+1}) \cdot Q_{i_t i_{t+1}}.$$

In this chapter, MDP means MCDP or MDCP.

Both types of MDPs may consider finite or infinite processes. For infinite processes, one typically assumes a discount factor $\gamma < 1$ and the reward of an infinite run is defined as $\sum_{t \ge 0} \gamma^t r_t$, where $r_t$ is the reward at time $t$. As mentioned earlier, discounts make little sense in many online stochastic optimizations. In online vehicle routing, customers placing requests later in the day should not be discounted. Similarly, late reservations should not be discounted in most applications. For this reason, we focus on finite time-horizon processes and the definition of an MDP $\mathcal{M_A}$ is assumed to contain a horizon $h \in \mathbb{N}$. Then the state set naturally is partitioned in $S = \bigcup_{0 \leq t \leq h} S_t$ and all transitions are from $S_t$ to $S_{t+1}$.

When $\mathcal{R}$ is a run, $w(\mathcal{R})$ denotes its reward, that is, the sum of its rewards for all times $t$ in $[0..h-1]$. A pair $(\mathcal{M_A}, \pi)$, where $\mathcal{M_A}$ is a MDP and $\pi$ a $\mathcal{M_A}$-policy, induces a probability distribution on the set of runs of $\mathcal{M_A}$ of length $h$. Since runs are associated with real-valued rewards, $(\mathcal{M_A}, \pi)$ also induces a reward probability distribution on $\mathbb{R}$, denoted by $\mathrm{rwd}(\mathcal{M_A}, \pi)$. The support of this distribution is finite since the set of all runs is finite.

## 13.3   Equivalence of MDCPs and MCDPs

This section shows that MDCPs and MCDPs have the same expressive power.

**Definition 13.5 (Simulation)** Let $\mathcal{M_A}$ and $\mathcal{M_B}$ be two MDPs. $\mathcal{M_B}$ simulates $\mathcal{M_A}$, denoted by $\mathcal{M_A} \preceq \mathcal{M_B}$, if there exists a map $\varphi : \mathrm{pol}(\mathcal{M_A}) \to \mathrm{pol}(\mathcal{M_B})$ that preserves the reward probability distribution, that is,

$$\forall \pi \in \mathrm{pol}(\mathcal{M_A}) : \ \mathrm{rwd}(\mathcal{M_A}, \pi) = \mathrm{rwd}(\mathcal{M_B}, \varphi(\pi))$$

The simulation relation is a preorder and induces an equivalence relation.

**Definition 13.6 (Equivalence of MDPs)** Let $\mathcal{M_A}$ and $\mathcal{M_B}$ be two MDPs. $\mathcal{M_A}$ is equivalent to $\mathcal{M_B}$, denoted by $\mathcal{M_A} \sim \mathcal{M_B}$, if $\mathcal{M_A} \preceq \mathcal{M_B}$ and $\mathcal{M_B} \preceq \mathcal{M_A}$.

Observe that $\mathcal{M_A} \preceq \mathcal{M_B}$ implies $\mathrm{rwd}(\mathcal{M_A}, \mathrm{pol}(\mathcal{M_A})) \subseteq \mathrm{rwd}(\mathcal{M_B}, \mathrm{pol}(\mathcal{M_B}))$ and $\mathcal{M_A} \sim \mathcal{M_B}$ implies $\mathrm{rwd}(\mathcal{M_A}, \mathrm{pol}(\mathcal{M_A})) = \mathrm{rwd}(\mathcal{M_B}, \mathrm{pol}(\mathcal{M_B}))$. Hence, for a decider, the two MDPs are equivalent, since they have the same set of reward distributions that are obtained by some policy. Note that $\mathcal{M_A} \sim \mathcal{M_B}$ does not imply the existence of a bijection between $\mathrm{pol}(\mathcal{M_A})$ and $\mathrm{pol}(\mathcal{M_B})$, since several policies can have the same reward distribution.

We are now in position to show that every MDCP is equivalent to some MCDP. The proof is based on two ideas. First, the MCDP works on a horizon of size $2h$, alternating decision-making steps and observations. Second, the MCDP must outsource the endogenous uncertainty of the MDCP. Since the MCDP does not know the state of the MDCP, the random inputs of the MCDP must be functions that take a state and an action as inputs and return a state. These functions have several degrees of freedom that make the proof technical.

**LEMMA 13.1**   Every MDCP is equivalent to some MCDP.

PROOF:   Let $\mathcal{M_A} = (S, s_0, A, T, R, h)$ be a finite horizon MDCP. We define an MCDP $\mathcal{M_B} = (S', s_0', \Xi, T', R', h')$ as follows:

- $h' = 2h$;
- $S' = S \times (\{\bot\} \cup A)$   (with $\bot \notin A$);
- $s'_0 = (s_0, \bot)$;
- $\Xi$ is defined by a tuple $(I, Q, \mu)$, where $I = S^{(S \times A)}$, that is, $I$ is the (finite) set of functions from $S \times A$ to $S$, $Q$ is the matrix where all elements are equal, and $\mu$ is defined by

$$\forall i \in S^{(S \times A)}, \quad \mu(\{i\}) = \prod_{(s,a) \in S \times A} T(s,a)(\{i(s,a)\});$$

- $T'$ is specified by

$$\begin{array}{ll} \forall s \in S, \ \forall i \in I, & T'((s, \bot), i) = \{s\} \times A \\ \forall s \in S, \ \forall a \in A, \ \forall i \in I, & T'((s, a), i) = \{(i(s,a), \bot)\}. \end{array}$$

- $R'$ is specified by

$$\begin{array}{ll} \forall s \in S, \ \forall i \in I, \ \forall s \in S', & R'((s, \bot), i, s') = 0 \\ \forall s \in S, \ \forall a \in A, \ \forall i \in I, & R'((s, a), i) = R(s, a, i(s, a)). \end{array}$$

The state of a process $\mathcal{M_B}$ is necessarily in $S \times \{\bot\}$ at date $2t$ and in $S \times A$ at date $2t+1$.

We now construct a map $\varphi$ from policies of $\mathcal{M_A}$ to policies of $\mathcal{M_B}$. Let $\pi$ be a $\mathcal{M_A}$-policy. On odd dates, $T'$ maps to a singleton and hence $\varphi(\pi)((s, a), i)$ must map to $i(s, a)$ with probability one. On even dates, transitions are deterministic since $T'((s, \bot), i)$ does not depend on $i$. For all $s \in S$ and $i \in I$, $\varphi(\pi)((s, \bot), i)$ is defined as

$$\begin{cases} \varphi(\pi)((s, \bot), i)(\{(s, a)\}) = \pi(s)(a) \\ \varphi(\pi)((s, \bot), i)(X) = 0 \text{ if } X \cap \{s\} \times A = \emptyset. \end{cases}$$

We now prove that $\varphi(\pi)$ has the same reward distribution as $\pi$. Consider the $\mathcal{M_A}$-run $\mathcal{R}$

$$\mathcal{R} = \left\langle s_0 \xrightarrow{a_0} s_1 \xrightarrow{a_1} \cdots \xrightarrow{a_{h-1}} s_h \right\rangle.$$

We must prove that there exists a set $X$ of $\mathcal{M_B}$-runs such that all the runs of $X$ have the same reward as $\mathcal{R}$ and the probability of $X$ under $(\mathcal{M_B}, \varphi(\pi))$ is the probability of $\mathcal{R}$ under $(\mathcal{M_A}, \pi)$. We define $X$ as the set of all valid runs of the form

$$\left\langle (s_0, \bot) \xrightarrow{i_0} (s_0, a_0) \xrightarrow{i_1} (s_1, \bot) \xrightarrow{i_2} (s_1, a_1) \xrightarrow{i_3} \cdots \xrightarrow{i_{2h-2}} (s_{h-1}, a_{h-1}) \xrightarrow{i_{2h-1}} (s_h, \bot) \right\rangle.$$

Such a run is valid if and only if $\forall t : i_{2t+1}(s_t, a_t) = s_{t+1}$. Observe that the set $X$ is characterized by two degrees of freedom: the $i_{2t}$'s on even dates do not matter and the value of $i_{2t+1}$ is significant only on the pair $(s_t, a_t)$. Since all elements of $Q$ are equal, we have

$$\forall t, \Pr(\xi_{t+1} = i_{t+1} \mid \xi_t = i_t) = \mu(i_{t+1}).$$

Thus each run in $X$ under $(\mathcal{M}_\mathcal{B}, \varphi(\pi))$ has probability

$$\prod_{0 \le t < h} \pi(s_t)(a)\mu(i_{2t}) \cdot \prod_{0 \le t < h} \mu(i_{2t+1}).$$

The probability of the set $X$ is the sum of these probabilities:

$$\sum_{i_{2t}} \sum_{\substack{i_{2t+1} \\ i_{2t+1}(s_t, a_t) = s_{t+1}}} \left( \prod_{0 \le t < h} \pi(s_t)(a)\mu(i_{2t}) \cdot \prod_{0 \le t < h} \mu(i_{2t+1}) \right),$$

which, by definition of $\mu$, is equal to

$$\prod_{0 \le t < h} \pi(s_t)(a) \cdot \prod_{0 \le t < h} T(s_t, a_t)(s_{t+1}).$$

This last formula is the probability of $\mathcal{R}$ under $(\mathcal{M}_\mathcal{A}, \pi)$. A similar reasoning applies to all $\mathcal{M}_\mathcal{A}$-runs, proving that

$$\forall x \in \mathbb{R} : \mathrm{rwd}(\mathcal{M}_\mathcal{A}, \pi)(x) \le \mathrm{rwd}(\mathcal{M}_\mathcal{B}, \varphi(\pi)).$$

As the support of these two distributions is finite and both are normalized, all these inequalities are in fact equalities.                                                                                  □

An MCDP $\mathcal{M}_\mathcal{A}$ can also be simulated by some MDCP $\mathcal{M}_\mathcal{B}$. The proof in-sources the exogenous uncertainty by considering states in $\mathcal{M}_\mathcal{B}$ that are the product of the states and inputs in $\mathcal{M}_\mathcal{A}$.

LEMMA 13.2  Every MCDP is equivalent to an MDCP.

PROOF:   We simply sketch the construction here. Let $\mathcal{M}_\mathcal{A} = (S, s_0, (I, Q, \mu), T, R, h)$ be an MCDP. We define an MDCP $\mathcal{M}_\mathcal{B} = (S', s_0', A, T', R', h')$ as follows:

- $h' = h + 1$;
- $S' = S \times I \cup \{(s_0, \bot)\}$;
- $s_0' = (s_0, \bot)$;
- $A = S$;

- To define $T'$, order $A = \{a_0, \ldots, a_{k-1}\}$ arbitrarily and denote by $a^X$ the element of $X \subseteq A$ with the smallest index and by $\sigma^X : A \to A$ the function defined as $\sigma^X(a) = a$ if $a \in X$, and $\sigma^X(a) = a^X$ otherwise. $T'$ can now be defined as follows. First, for all $a \in A$, $T'((s_0, \perp), a)$ is the probability distribution defined by

$$T'((s_0, \perp), a)(s_0, i) = \mu(i),$$

all other elements of $S'$ having probability zero. Second,

$$T'((x, i_1), a)(y, i_2) = \begin{cases} Q_{i_1, i_2} & \text{if } y = \sigma^{T(x, i_1)}(a) \\ 0 & \text{otherwise.} \end{cases}$$

- $R'$ is specified by

$$R'((s_0, \perp), a, s') = 0.$$
$$R'((x, i_1), a, (y, i_2)) = R(x, a, y).$$

Finally define $\varphi : \text{pol}(\mathcal{M}_A) \to \text{pol}(\mathcal{M}_B)$ as $\varphi(\pi)(s_0, \perp)(a) = \frac{1}{|A|}$ and $\varphi(\pi)(x, i) = \pi(x, i)$. $\square$
The proof also points out a fundamental drawback of MDCPs for online stochastic combinatorial optimization: the number of states in the MDCPs is now the product of the number of states and the number of inputs in the MCDPs. This is unavoidable since policies in MCDPs are defined over pairs $(s_t, i_t)$.

## 13.4 Online Anticipatory Algorithms

Although MCDPs and MDCPs have the same expressive power, MCDPs are important modeling tools for computational reasons. They allow the design of online anticipatory algorithms that exploit anticipatory relaxations and, under some anticipativity assumption, make near-optimal decisions quickly. It is important to understand the difference between MCDPs and MDCPs. Recall that the goal in MDPs is to maximize the expected reward. In MDCPs, from state $s_0$, we must compute

$$\max_{a_0} \mathbb{E}_{s_1} \left[ R(s_0, a_0, s_1) + \max_{a_1} \mathbb{E}_{s_2} (\ldots \mid a_1, s_1) \mid a_0, s_0 \right].$$

For simplicity, all expectations are implicitly conditional to the past in the following and $\mathcal{R}$ denotes the run defined by past decisions and random variables. Since $R(s_t, a_t, s_{t+1})$ does not depend on the future, this is equivalent to

$$\max_{a_0} \mathbb{E}_{s_1} \left[ \max_{a_1} \mathbb{E}_{s_2} \left( \ldots \mathbb{E}_{s_h} w(\langle s_0 \xrightarrow{a_0} s_1 \xrightarrow{a_1} \cdots \xrightarrow{a_{h-1}} s_h \rangle) \right) \right] \tag{13.4.1}$$

where $\mathcal{R}$ is the run defined by the pair $(a_i, s_{i+1})$. In MCDPs, given state $s_0$ and observations $i_0$, we must compute

$$\max_{s_1} \mathbb{E}_{\xi_1} \left[ \max_{s_2} \mathbb{E}_{\xi_2} \left( \ldots \mathbb{E}_{s_h} (w(\langle s_0 \xrightarrow{i_0} s_1 \xrightarrow{\xi_1} \cdots \xrightarrow{\xi_{h-1}} s_h \rangle)) \right) \right]. \qquad (13.4.2)$$

Formulas (13.4.1) and (13.4.2) are not as similar as they seem. In (13.4.1), the distribution of $s_{t+1}$ depends on both $s_t$ and $a_t$. In (13.4.2), the input process is exogenous and $\xi_{t+1}$ depends only on $\xi_t$, not on the past decisions. Consider now what happens when $\mathbb{E}$ and max are commuted. For (13.4.1), the formula

$$\max_{a_0} \mathbb{E}_{s_1} \mathbb{E}_{s_2} \ldots \mathbb{E}_{s_h} \left( \max_{a_1} \ldots \max_{a_{h-1}} w(\langle s_0 \xrightarrow{a_0} s_1 \xrightarrow{a_1} \cdots \xrightarrow{a_{h-1}} s_h \rangle) \right) \qquad (13.4.3)$$

makes no sense while, for (13.4.2), the formula

$$\max_{s_1} \mathbb{E}_{\xi_1} \mathbb{E}_{\xi_2} \ldots \mathbb{E}_{\xi_{h-1}} \left( \max_{s_2} \ldots \max_{s_h} w(\langle s_0 \xrightarrow{i_0} s_1 \xrightarrow{\xi_1} \cdots \xrightarrow{\xi_{h-1}} s_h \rangle) \right) \qquad (13.4.4)$$

is well defined. It is precisely the *anticipatory relaxation* that provides an upper bound to (13.4.2). Observe that the expression

$$\max_{s_2} \ldots \max_{s_h} w(\langle s_0 \xrightarrow{i_0} s_1 \xrightarrow{\xi_1} \cdots \xrightarrow{\xi_{h-1}} s_h \rangle)$$

is a deterministic optimization since all the uncertainty $\xi_{1..h-1}$ has been revealed. This means that every MCDP has a natural underlying deterministic optimization problem that can be solved by dedicated optimization techniques. In the following, $\mathcal{O}$ denotes such an optimization algorithm: given a state $s_t$ and an input sequence $i_{t..h-1}$, $\mathcal{O}(s_t, i_{t..h-1})$ returns a run

$$\mathcal{R} = \langle s_t \xrightarrow{i_t} s_{t+1} \xrightarrow{i_{t+1}} \cdots \xrightarrow{i_{h-1}} s_h \rangle$$

maximizing $w(\mathcal{R})$. This algorithm may be a dynamic program in packet scheduling, an integer program in online reservations, and a constraint program in vehicle routing. Now (13.4.4) becomes

$$\max_{s_1 \in T(s_0, i_0)} \mathbb{E}_{\xi_{1..h-1}} w(s_0 \xrightarrow{i_0} \mathcal{O}(s_1, \xi_{1..h-1})).$$

We now define the concept of *online anticipatory algorithm* for MCDPs. Online anticipatory algorithms make decisions online as the uncertainty is revealed using the anticipatory relaxation.

ONLINE ALGORITHM $\mathcal{E}(\langle i_0, \ldots, i_{h-1} \rangle)$
1  $\sigma_0 \leftarrow s_0$;
2  **for** $t \in 0..h-1$ **do**
3    $\sigma_{t+1} \leftarrow \mathcal{E}(\sigma_t, i_t)$;
4  **return** $\gamma_h$;


ALGORITHM $\mathcal{E}(\sigma_t, i_t)$
1  **for** $\sigma \in T(\sigma_t, i_t)$ **do**
2    $f(\sigma) \leftarrow R(\sigma_t, i_t, \sigma)$;
3  **for** $k \leftarrow 1 \ldots m$ **do**
4    $i_{t+1..h-1}^k \leftarrow \text{SAMPLE}(t+1, h-1)$;
5    **for** $\sigma \in T(\sigma_t, i_t)$ **do**
6      $f(\sigma) \leftarrow f(\sigma) + w(\mathcal{O}(\sigma, i_{t+1..h-1}^k))$;
7  **return** arg-max$(\sigma \in T(\sigma_t, i_t))$ $f(\sigma)/m$;

**Figure 13.1:** The Online Anticipatory Algorithm $\mathcal{E}$ for MCDPs.

**Definition 13.7 (Online Anticipatory Algorithm)** An online anticipatory algorithm $\mathcal{A}$ produces a sequence of online decisions

$$s_1 = \mathcal{A}(s_0, i_0), \ldots, s_h = \mathcal{A}(s_{h-1}, i_{h-1})$$

using approximations of the form

$$\mathbb{E}_{\xi_{t..h-1}} w(\mathcal{O}(s_t, \xi_{t..h-1})) \tag{13.4.5}$$

for

$$\mathbb{E}_{\xi_t} \left[ \max_{s_{t+1}} \mathbb{E}_{\xi_{t+1}} \left( \ldots \max_{s_h} w(\langle s_t \xrightarrow{\xi_t} s_{t+1} \xrightarrow{\xi_{t+1}} \cdots \xrightarrow{\xi_{h-1}} s_h \rangle) \right) \right]. \tag{13.4.6}$$

Figure 13.1 depicts the anticipatory algorithm $\mathcal{E}$ for MCDPs. The first four lines specify the online decision process and indicate that, at time $t$, algorithm $\mathcal{E}$ must choose a new state $\sigma_{t+1}$, given the current state $\sigma_t$ and the revealed input $i_t$. The last seven lines specify how algorithm $\mathcal{E}$ takes the decision at time $t$. $\mathcal{E}$ uses $m$ scenarios and initializes the scores of all possible decisions in lines 1 and 2 using the reward of moving from $\sigma_t$ to $\sigma \in T(\sigma_t, i_t)$. It then generates $m$ scenarios (lines 3 and 4), evaluates each possible decision on each scenario and increments the score of the decisions accordingly (lines 5 and 6). It returns the decision with the highest score in line 7. Once again,

$\mathcal{E}$ is appropriate when the number of possible decisions $|T(s_t, i_t)|$ is not too large. The consensus and regret algorithms can be derived in a similar fashion for handling time constraints and the case when $|T(s_t, i_t)|$ is large.

## 13.5 The Approximation Theorem for Anticipative MCDPs

This section studies the expected loss of algorithm $\mathcal{E}$ with respect to the optimal, a posteriori solution, that is, the expected difference in reward between decisions made by $\mathcal{E}$ and decisions taken by an optimization algorithm when the inputs are fully revealed. Its main result is to show that the expected loss of $\mathcal{E}$, under the $\epsilon$-anticipativity assumption, is $O(h\epsilon)$ using a polynomial number of optimizations at each step.

Algorithm $\mathcal{E}$ makes two kinds of error at each step. First, it uses the anticipatory relaxation

$$\mathop{\mathbb{E}}_{\xi_{t+1..h-1}} w(\mathcal{O}(s_{t+1}, \xi_{t+1..h-1})) \tag{13.5.7}$$

instead of

$$\mathop{\mathbb{E}}_{\xi_{t+1}} \left[ \max_{s_{t+2}} \mathop{\mathbb{E}}_{\xi_{t+2}} \left( \ldots \max_{s_h} w(\langle s_{t+1} \xrightarrow{\xi_{t+1}} s_{t+2} \xrightarrow{\xi_{t+2}} \cdots \xrightarrow{\xi_{h-1}} s_h \rangle) \right) \right]. \tag{13.5.8}$$

Second, it approximates (13.5.7) by sampling, that is,

$$\frac{1}{m} \sum_{k=1}^{m} w(\mathcal{O}(s_{t+1}, i_{t+1..h-1}^k)) \tag{13.5.9}$$

where $i_{t+1..h-1}^k$ is a realization of $\xi_{t+1..h-1}$ $(1 \le k \le m)$. The proof relies on the concept of local loss that captures the loss of a suboptimal decision at step $t$ assuming that all subsequent decisions are optimal. It proceeds in three steps:

1. Lemma 13.4 first shows that the expected global loss is the sum of the expected local losses at each step.

2. Lemma 13.5 then bounds the sampling error, that is, the loss induced by using (13.5.9) instead of (13.5.7).

3. Finally, theorem 13.1 shows that, under the $\epsilon$-anticipativity assumption, the local expected losses are small and so is the expected loss of algorithm $\mathcal{E}$.

As before, $\xi_{0...h-1}$ is the sequence of random variables of the input process. Moreover, $\sigma_{1...h}$ denotes the sequence of random variables representing the states returned by the algorithm $\mathcal{E}$, that is

$$\begin{aligned} \sigma_1 &= \mathcal{E}(\sigma_0, i_0) \\ \sigma_2 &= \mathcal{E}(\sigma_1, i_1) \\ &\cdots \\ \sigma_h &= \mathcal{E}(\sigma_{h-1}, i_{h-1}) \end{aligned}$$

where $\sigma_0 = s_0$. The state $\sigma_t$ is a random variable that is a function of the realization $i_t$ of the random variable $\xi_t$ (that is, the input revealed at time $t$), $\sigma_{t-1}$, and the $m$ samples used in $\mathcal{E}$. Indeed different samples may induce different choices for $\sigma_{t+1}$ given the same pair $(\sigma_t, i_t)$ as should be clear from the experimental results presented in this book. Consider the random run

$$\mathcal{R} = \langle \sigma_0 \xrightarrow{i_0} \cdots \xrightarrow{i_{h-1}} \sigma_h \rangle.$$

We use $\mathcal{R}_t$ to denote the run

$$\mathcal{R}_t = \langle \sigma_0 \xrightarrow{i_0} \cdots \sigma_{t-1} \xrightarrow{i_{t-1}} \mathcal{O}(\sigma_t, i_{t...h-1}) \rangle,$$

which coincides with $\mathcal{R}$ until reaching state $\sigma_t$, but uses the optimization algorithm on the sequence $i_{t...h-1}$ thereafter. Note that $\mathcal{R} = \mathcal{R}_h$ and that the goal is to study

$$\mathop{\mathbb{E}}_{\mathcal{R}} \left[ w(\mathcal{R}_0) - w(\mathcal{R}) \right].$$

The following definition reframes the concept of local loss and expected local loss in the context of MCDPs. The expected local loss captures the expected loss induced by a decision at time $t$ assuming that all subsequent decisions are taken optimally.

**Definition 13.8 (Local Loss and Expected Local Loss)** Let $s_t \in S_t$, $i_{t..h-1}$ be a scenario, and $s_{t+1} \in T(s_t, i_t)$. The local loss of $s_{t+1}$ wrt $s_t$ and $i_{t..h-1}$, denoted by $\delta(s_t, i_{t..h-1}, s_{t+1})$, is defined as

$$\delta(s_t, i_{t..h-1}, s_{t+1}) = w(\mathcal{O}(s_t, i_{t...h-1})) - w(s_t \xrightarrow{i_t} \mathcal{O}(s_{t+1}, i_{t+1...h-1})).$$

The expected local loss of $s_{t+1}$ with respect of $s_t$ and $i_t$, denoted by $\Delta(s_t, i_t, s_{t+1})$ is defined as

$$\Delta(s_t, i_t, s_{t+1}) = \mathop{\mathbb{E}}_{\xi_{t+1..h-1}} \delta(s_t, i_t : \xi_{t+1..h-1}, s_{t+1}).$$

Observe that, in the definition, the optimal decision at time $t$ is computed by algorithm $\mathcal{O}$ and that it may differ from scenario to scenario. In contrast, the decision $s_{t+1}$ is the same for all scenarios. It is useful to separate the two errors occurring in algorithm $\mathcal{E}$. The first error is the anticipatory gap due to the use of the anticipatory relaxation. Intuitively, the anticipatory gap is the price of online computation: at a time $t$, the decision for all scenarios must be the same for all scenarios.

**Definition 13.9 (The Anticipatory Gap)** Let $s_t \in S_t$ and $i_t \in I$. The *anticipatory gap* of $(s_t, i_t)$, denoted by $\Delta_g(s_t, i_t)$ is defined as

$$\Delta_g(s_t, i_t) = \mathop{\mathbb{E}}_{\xi_{t+1...h-1}} w(\mathcal{O}(s_t, i_t : \xi_{t+1...h-1}) - \max_{s^\star_{t+1} \in T(s_t, i_t)} \mathop{\mathbb{E}}_{\xi_{t+1...h-1}} w(s_t \xrightarrow{i_t} \mathcal{O}(s^\star_{t+1}, \xi_{t+1...h-1})).$$

Second, algorithm $\mathcal{E}$ uses sampling to approximate

$$\max_{s_{t+1}^{\star} \in T(s_t, i_t)} \mathbb{E}_{\xi_{t+1\ldots h-1}} w(s_t \xrightarrow{i_t} \mathcal{O}(s_{t+1}^{\star}, \xi_{t+1\ldots h-1})).$$

**Definition 13.10 (Choice Error)** Let $s_t \in S_t$ and $i_t \in I$. The *choice error* of $s_{t+1}$ with respect to $(s_t, i_t)$, denoted by $\Delta_c(s_t, i_t, s_{t+1})$, is defined as

$$\max_{s_{t+1}^{\star} \in T(s_t, i_t)} \mathbb{E}_{\xi_{t+1\ldots h-1}} w(s_t \xrightarrow{i_t} \mathcal{O}(s_{t+1}^{\star}, \xi_{t+1\ldots h-1})) - \mathbb{E}_{\xi_{t+1\ldots h-1}} w(s_t \xrightarrow{i_t} \mathcal{O}(s_{t+1}, \xi_{t+1\ldots h-1})).$$

The expected local loss at time $t$ for algorithm $\mathcal{E}$ is the summation of these two errors.

LEMMA 13.3 (EXPECTED LOCAL LOSS) Let $s_t \in S_t$ and $i_t \in I$. The *expected local error* $s_{t+1}$ with respect to $(s_t, i_t)$ satisfies

$$\Delta(s_t, i_t, s_{t+1}) = \Delta_g(s_t, i_t) + \Delta_c(s_t, i_t).$$

Observe that all the $\Delta$s are nonnegative. Moreover,

$$\mathbb{E}_{\xi_{t+1\ldots h-1}} [w(\mathcal{R}_t) - w(\mathcal{R}_{t+1})] = \mathbb{E}_{\xi_{t+1\ldots h-1}} [w(\mathcal{O}(\sigma_t, \xi_{t+1\ldots h-1}) - w(\sigma_t \xrightarrow{i_t} \mathcal{O}(\sigma_{t+1}, \xi_{t+1\ldots h-1})))]$$

and thus

$$\mathbb{E}_{\xi_{t+1\ldots h-1}} [w(\mathcal{R}_t) - w(\mathcal{R}_{t+1})] = \Delta(\sigma_t, i_t, \sigma_{t+1}). \tag{13.5.10}$$

We are thus in position to present the first lemma that shows that the global loss of algorithm $\mathcal{E}$ is the sum of its local expected losses.

LEMMA 13.4 (SUMMATION OF EXPECTED LOCAL LOSSES)

$$\mathbb{E}_{\mathcal{R}}[w(\mathcal{R}_0) - w(\mathcal{R})] = \sum_{t=0}^{h-1} \mathbb{E}_{\substack{\sigma_t, \xi_t, \\ \sigma_{t+1}}} \Delta(\sigma_t, \xi_t, \sigma_{t+1})$$

PROOF:   Define

$$A_t = \mathbb{E}_{\mathcal{R}}[w(\mathcal{R}_t) - w(\mathcal{R})].$$

The goal is to compute $A_0$. As $\mathcal{R} = \mathcal{R}_h$, we have $A_h = 0$ and $A_0$ can be computed inductively. Let

$0 \leq t < h$. Then,

$$
\begin{aligned}
A_t &= \mathop{\mathbb{E}}_{\mathcal{R}}[w(\mathcal{R}_t) - w(\mathcal{R})] \\
&= \mathop{\mathbb{E}}_{\mathcal{R}}[w(\mathcal{R}_t) - w(\mathcal{R}_{t+1}) + w(\mathcal{R}_{t+1}) - w(\mathcal{R})] \\
&= \mathop{\mathbb{E}}_{\substack{\xi_{0\ldots t} \\ \sigma_{0\ldots t}}} \left( \mathop{\mathbb{E}}_{\substack{\xi_{t+1\ldots h-1} \\ \sigma_{t+1\ldots h}}} [w(\mathcal{R}_t) - w(\mathcal{R}_{t+1})] + \mathop{\mathbb{E}}_{\substack{\xi_{t+1\ldots h-1} \\ \sigma_{t+1\ldots h}}} [w(\mathcal{R}_{t+1}) - w(\mathcal{R})] \right).
\end{aligned}
$$

Now, as $\mathcal{R}_t$ and $\mathcal{R}_{t+1}$ do not depend on $\sigma_{t+2\ldots h}$, we have

$$
A_t = \mathop{\mathbb{E}}_{\substack{\xi_{0\ldots t} \\ \sigma_{0\ldots t}}} \left( \mathop{\mathbb{E}}_{\sigma_{t+1}} \mathop{\mathbb{E}}_{\xi_{t+1\ldots h-1}} [w(\mathcal{R}_t) - w(\mathcal{R}_{t+1})] + \mathop{\mathbb{E}}_{\substack{\xi_{t+1\ldots h-1} \\ \sigma_{t+1\ldots h}}} [w(\mathcal{R}_{t+1}) - w(\mathcal{R})] \right),
$$

which, by (13.5.10), leads to

$$
A_t = \mathop{\mathbb{E}}_{\substack{\xi_{0\ldots t} \\ \sigma_{0\ldots t}}} \left( \mathop{\mathbb{E}}_{\sigma_{t+1}} \Delta(\sigma_t, \xi_t, \sigma_{t+1}) + \mathop{\mathbb{E}}_{\substack{\xi_{t+1\ldots h-1} \\ \sigma_{t+1\ldots h}}} [w(\mathcal{R}_{t+1}) - w(\mathcal{R})] \right).
$$

By linearity of expectations,

$$
A_t = \mathop{\mathbb{E}}_{\substack{\xi_{0\ldots t} \\ \sigma_{0\ldots t+1}}} [\Delta(\sigma_t, \xi_t, \sigma_{t+1})] + \mathop{\mathbb{E}}_{\substack{\xi_{0\ldots h-1} \\ \sigma_{0\ldots h}}} [w(\mathcal{R}_{t+1}) - w(\mathcal{R})].
$$

As $\Delta(\sigma_t, \xi_t, \sigma_{t+1})$ depends neither on $\xi_{0\ldots t-1}$, nor on $\sigma_{0\ldots\sigma_{t-1}}$, we have

$$
A_t = \mathop{\mathbb{E}}_{\substack{\sigma_t, \xi_t, \\ \sigma_{t+1}}} [\Delta(\sigma_t, \xi_t, \sigma_{t+1})] + \mathop{\mathbb{E}}_{\substack{\xi_{0\ldots h-1} \\ \sigma_{0\ldots h}}} [w(\mathcal{R}_{t+1}) - w(\mathcal{R})],
$$

which, by definition of $A_{t+1}$, gives

$$
A_t = \mathop{\mathbb{E}}_{\substack{\sigma_t, \xi_t, \\ \sigma_{t+1}}} [\Delta(\sigma_t, \xi_t, \sigma_{t+1})] + A_{t+1}
$$

and the result follows. $\qquad\square$

We are now ready to bound the choice error in algorithm $\mathcal{E}$ at step $t$.

LEMMA 13.5 (CHOICE ERROR) Let $s \in S$, $i \in I$, and $\overline{s}$ be the random variable $\mathcal{E}(s, i)$. We have

$$
\mathop{\mathbb{E}}_{\overline{s}} \Delta_c(s, i, \overline{s}) \leq \sum_{s' \in T(s,i)} \Delta_c(s, i, s')\, e^{\frac{-m \Delta_c(s,i,s')^2}{2\sigma_{s,i,s'}^2}}
$$

where $\sigma_{s,i,s'}$ is the standard deviation (over $\xi_{t+1..h-1}$) of the local loss $\delta(s, i : \xi_{t+1..h-1}, s')$.

PROOF:   Let $t$ be such that $s \in S_t$. When called with arguments $(s,i)$, algorithm $\mathcal{E}$ generates $m$ scenarios $i_{t+1..h-1}^1, \ldots, i_{t+1..h-1}^m$ and computes, for each state $s' \in T(s,i)$, an approximation

$$f(s') = \frac{1}{m} \sum_{k=1}^{m} w\left(s \xrightarrow{i} \mathcal{O}(s', \xi_{t+1..h-1}^k)\right)$$

of the expectation

$$\mathop{\mathbb{E}}_{\xi_{t+1..h-1}} \left[ w(s \xrightarrow{i} \mathcal{O}(s', \xi_{t+1..h-1})) \right].$$

Algorithm $\mathcal{E}$ chooses state $\bar{s}$, that is,

$$\bar{s} = \operatorname*{arg\,max}_{s' \in \mathcal{S}(s'_{t-1}, i_t)} f(s').$$

Consider a state $s^\star$ such that $\Delta_c(s, i, s^\star) = 0$. Define $L_{t,s'}$ as

$$L_{t,s'} = (f(s') - f(s^\star)) - \left( \mathop{\mathbb{E}}_{\xi_{t+1..h-1}} w(s \xrightarrow{i} \mathcal{O}(s', \xi_{t+1..h-1})) - \mathop{\mathbb{E}}_{\xi_{t+1..h-1}} w(s \xrightarrow{i} \mathcal{O}(s^\star, \xi_{t+1..h-1})) \right).$$

By definition of the choice error and the optimality of $s^\star$,

$$\Delta_c(s, i, s') = \mathop{\mathbb{E}}_{\xi_{t+1..h-1}} \left[ w(s \xrightarrow{i} \mathcal{O}(s^\star, \xi_{t+1..h-1})) \right] - \mathop{\mathbb{E}}_{\xi_{t+1..h-1}} \left[ w(s \xrightarrow{i} \mathcal{O}(s', \xi_{t+1..h-1})) \right].$$

The condition $f(s') - f(s^\star) \geq 0$ implies

$$\Delta_c(s, i, s') \leq L_{t,s'} \tag{13.5.11}$$

and is thus a necessary condition for the selection of $s'$ by $\mathcal{E}$. In particular, since $f(\bar{s})$ is maximum, we have $f(\bar{s}) \geq f(s^\star)$. Moreover, the difference $f(s') - f(s_t^\star)$ is the average of $m$ independent, identically distributed, random variables, each with mean

$$\left( \mathop{\mathbb{E}}_{\xi_{t+1..h-1}} \left[ w(s \xrightarrow{i} \mathcal{O}(s', \xi_{t+1..h-1})) \right] - \mathop{\mathbb{E}}_{\xi_{t+1..h-1}} \left[ w(s \xrightarrow{i} \mathcal{O}(s^\star, \xi_{t+1..h-1})) \right] \right)$$

and standard deviations $\sigma_{s,i,s'}$. By the central limit theorem, we can argue that

$$\sqrt{m} L_{t,s'} / \sigma_{t,s'}$$

is a normal distribution $N(0,1)$. By definition of expectation,

$$\mathop{\mathbb{E}}_{\bar{s}} \Delta_c(s, i, \bar{s}) = \sum_{s' \in T(s,i)} \Delta_c(s, i, s') \Pr(\bar{s} = s').$$

By the necessary condition (13.5.11), it follows that

$$\Pr(\overline{s} = s') \le \Pr(\Delta_c(s, i, s') \le L_{t,s'})$$

and, by applying a Chernoff bound for the standard normal random variable,

$$\Pr(\Delta_c(s, i, s') \le L_{t,s'}) \le e^{-m\Delta_c(s,i,s')^2/2\sigma_{s,i,s'}^2}.$$

Hence,

$$
\begin{aligned}
\mathbb{E}_{\overline{s}}[\Delta_c(s, i, \overline{s})] &= \sum_{s' \in T(s,i)} \Delta_c(s, i, s') \Pr(\overline{s} = s') \\
&\le \sum_{s' \in T(s,i)} \Delta_c(s, i, s') \Pr(\Delta_c(s, i, s') \le L_{i,s'}) \\
&\le \sum_{s' \in T(s,i)} \Delta_c(s, i, s') e^{-m\Delta_c(s,i,s')^2/2\sigma_{t,s'}^2}.
\end{aligned}
$$

$\square$

Before proving the theorem, it is necessary to introduce the $\epsilon$-anticipativity assumption.

**Definition 13.11 ($\epsilon$-Anticipativity)** Let $s_t \in S_t$ and $i_t \in I$. An MCDP is $\epsilon$-anticipative if

$$\forall (s_t, i_t) \in S_t \times I : \quad \Delta_g(s_t, i_t) \le \epsilon.$$

The next section discusses this assumption in detail both experimentally and theoretically. The main theorem bounds the expected loss of $\mathcal{E}$.

THEOREM 13.1 (PERFORMANCE OF $\mathcal{E}$) Let $\mathcal{M}_A$ be an $\epsilon$-anticipative MCDP.

$$\mathbb{E}_{\mathcal{R}}[w(\mathcal{R}_0) - w(\mathcal{R})] \le h\left(\epsilon + \max_{\substack{s \in S \\ i \in I}} \sum_{s' \in T(s,i)} \Delta_c(s, i, s') e^{\frac{-m\Delta_c(s,i,s')^2}{2\sigma_{s,i,s'}^2}}\right).$$

PROOF: By definition of the local expected error, we have

$$\mathbb{E}_{\substack{\sigma_t, \xi_t, \\ \sigma_{t+1}}} \Delta(\sigma_t, \xi_t, \sigma_{t+1}) = \mathbb{E}_{\substack{\sigma_t, \xi_t, \\ \sigma_{t+1}}} \Delta_g(\sigma_t, i_t) + \mathbb{E}_{\substack{\sigma_t, \xi_t, \\ \sigma_{t+1}}} \Delta_c(\sigma_t, \xi_t, \sigma_{t+1}).$$

By the $\epsilon$-anticipativity assumption, it follows that

$$\mathbb{E}_{\substack{\sigma_t, \xi_t, \\ \sigma_{t+1}}} \Delta(\sigma_t, \xi_t, \sigma_{t+1}) \le \epsilon + \mathbb{E}_{\substack{\sigma_t, \xi_t, \\ \sigma_{t+1}}} \Delta_c(\sigma_t, \xi_t, \sigma_{t+1})$$

and hence, by lemma 13.5,

$$
\mathop{\mathbb{E}}_{\substack{\sigma_t,\,\xi_t,\\\sigma_{t+1}}} \Delta(\sigma_t,\xi_t,\sigma_{t+1}) \leq \epsilon + \mathop{\mathbb{E}}_{\sigma_t,\,\xi_t} \left[ \sum_{s'\in T(\sigma_t,\xi_t)} \Delta_c(\sigma_t,\xi_t,s')\, e^{\frac{-m\Delta_c(\sigma_t,\xi_t,s')^2}{2\sigma^2_{\sigma_t,\xi_t,s'}}} \right].
$$

By lemma 13.4,

$$
\mathop{\mathbb{E}}_{\mathcal{R}}[w(\mathcal{R}_0) - w(\mathcal{R})] = \sum_{t=0}^{h-1} \mathop{\mathbb{E}}_{\substack{\sigma_t,\,\xi_t,\\\sigma_{t+1}}} \Delta(\sigma_t,\xi_t,\sigma_{t+1})
$$

$$
\leq \sum_{t=0}^{h-1} \left( \epsilon + \max_{\substack{s\in S\\i\in I}} \sum_{s'\in T(s,i)} \Delta_c(s,i,s')\, e^{\frac{-m\Delta_c(s,i,s')^2}{2\sigma^2_{s,i,s'}}} \right)
$$

$$
\leq h \left( \epsilon + \max_{\substack{s\in S\\i\in I}} \sum_{s'\in T(s,i)} \Delta_c(s,i,s')\, e^{\frac{-m\Delta_c(s,i,s')^2}{2\sigma^2_{s,i,s'}}} \right). \qquad \square
$$

As a corollary, the expected error is $O(h\epsilon)$ when $m$ is $\Omega\big(\log\big(h\max_{(s,i)\in S\times I}|T(s,i)|\big)\big)$. Hence, algorithm $\mathcal{E}$ provides a near-optimal approximation of the offline, a posteriori problem with a polynomial number of optimizations.

## 13.6   The Anticipativity Assumption

This section discusses why the $\epsilon$-anticipativity assumption holds for many practical online applications. We first show how to verify whether an MCDP is $\epsilon$-anticipative experimentally.

LEMMA 13.6 (BOUNDING THE ANTICIPATIVITY GAP) Consider an MCDP $\mathcal{M}$, a state $s_t \in S_t$, an input $i_t$, and a state $s_{t+1} \in S_{t+1}$. If the probability that a run starting with $s_t \xrightarrow{i_t} s_{t+1}$ be optimal for a scenario $i_{t+1..h-1}$ is at least $p$, then

$$
\Delta_g(s_t,i_t) \leq (1-p)\delta_{\max}
$$

where $\delta_{\max} = \max_{i_{t+1..h-1}} \delta(s_t,i_{t..h-1},s_{t+1})$.

PROOF:   Consider all the possible realizations

$$
i^1_{t+1..h-1},\ldots,i^k_{t+1..h-1}
$$

of $\xi_{t+1..h-1}$ conditioned to $\xi_t = i_t$. Define

$$
P = \{i_t : i^j_{t+1..h-1} \mid 1 \leq j \leq k\}
$$

and

$$A = \{i_{t..h-1} \in P \mid \delta(s_t, i_{t..h-1}, s_{t+1}) = 0\}.$$

By definition of the anticipatory gap,

$$\Delta_g(s_t, i_t) \leq \mathop{\mathbb{E}}_{\xi_{t+1..h-1}} \delta(s_t, i_t : \xi_{t+1..h-1}, s_{t+1})$$

$$= \Pr(A) \cdot \mathop{\mathbb{E}}_{\xi_{t+1..h-1}} \left( \delta(s_t, \xi_{t..h-1}, s_{t+1}) \mid \xi_{t..h-1} \in A \right)$$

$$+ \Pr(\overline{A}) \cdot \mathop{\mathbb{E}}_{\xi_{t+1..h-1}} \left( \delta(s_t, \xi_{t..h-1}, s_{t+1}) \mid \xi_{t..h-1} \notin A \right)$$

$$= \Pr(\overline{A}) \cdot \mathop{\mathbb{E}}_{\xi_{t+1..h-1}} \left( \delta(s_t, \xi_{t..h-1}, s_{t+1} \mid \xi_{t..h-1} \notin A \right)$$

$$\leq (1 - \Pr(A)) \cdot \max_{i_{t+1...h-1}} \delta(s_t, i_{t..h-1}, s_{t+1}).$$

$\square$

The lemma gives a way to bound the anticipation error by approximating the probability that the selected decision $s_{t+1}$ be optimal in the scenarios. This is exactly the experimental data collected on the various applications and the probability turns out to be high on all of them. Figures 13.2, 13.3, and 13.4 depict again the experimental results on $\epsilon$-anticipativity on three applications of increasing complexity: the online packet scheduling problem, the online multiknapsacks with deadlines, and online multiple-vehicle routings with time windows. Figures 13.2 and 13.3 depict the percentage of scenarios whose optimal solutions agree with the selected choice as a function of the number of scenarios. Figure 13.4 plots the same information as a function of time $t$. Observe that the probability $p$ is around 0.9 in the average for online packet scheduling, 0.8 for online multiknapsack, and 0.75 for online vehicle routing. The value $\delta_{\max}$ is bounded easily in packet scheduling and multiknapsack: it is the difference between the most valuable and least valuable requests. It is harder to bound in vehicle routing due to vehicle traveling times and time windows. Experimentally, it is very small as well.

Why is the anticipatory gap small on these applications? The intuition is that there is a temporal component in these applications that keeps the anticipatory gap small. In packet scheduling, the online algorithm has a tendency to serve older packets since they expire earlier: it is only when a packet is very valuable that it is selected immediately. Moreover, the impact of selecting such an older packet on the next couple of decisions does not significantly depend on the order of arriving packets. In vehicle routing, the time windows during which a customer can be served are necessarily later than the request time, implying once again that the decision at a time $t$ will concern older requests that are available to all scenarios.

One may wonder why there are several online anticipatory algorithms for vehicle routing, while there is only a single algorithm on MCDPs. The main issue in online vehicle routing is where the vehicle is allowed to move — either to known customers only, to known customers and the current

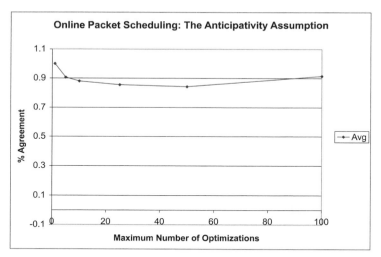

**Figure 13.2:** $\epsilon$-Anticipativity in Online Packet Scheduling.

location, or to any location. In the context of MCDPs, these strategies correspond to a different choice of states and are now in the realm of modeling instead of being an algorithmic decision. Such modeling issues in online stochastic combinatorial optimization are likely to abound and this book has barely touched on this topic.

## 13.7   Beyond Anticipativity

Not all problems will be $\epsilon$-anticipative but the concept of MCDPs may still be useful in these circumstances. As mentioned in chapter 3, it is possible to generalize the algorithms when the $\epsilon$-anticipativity assumption does not hold. Figure 13.5 depicts algorithm $\mathcal{E}^h$, which explores a tree of depth $h - t + 1$ to make the decision at time $t$. Algorithm $\mathcal{E}^h(\sigma_t, i_t)$ simply selects the successor state $\sigma_{t+1} \in T(\sigma_t, i_t)$ with the best evaluation. The function EvalState-$\mathcal{E}^h(t, \sigma_t)$ evaluates the state $\sigma_t$ at time $t$. If $t = h$, the evaluation is complete and the function simply returns the objective value of the state (lines 1 and 2). Otherwise, the function generates $m$ samples of $\xi_t$ (line 6), evaluates the successors of $\sigma_{t+1}$ for each realization $i_t$, and increments $e$ with the best evaluation of the successors in $T(\sigma_t, i_t)$ (line 7). The function returns the value $e/m$, where $m$ is the number of samples.

It is possible to relate the quality of algorithm $\mathcal{E}^h$ to the optimal value $w_0^*(s_0 = \sigma_0)$ of the MCDPs where

$$
\begin{aligned}
w_h^*(\sigma_h) &= 0; \\
w_t^*(\sigma_t) &= \mathop{\mathbb{E}}_{\xi_t} \left[ \max_{\sigma_{t+1} \in T(\sigma_t, \xi_t)} R(\sigma_t, \xi_t, \sigma_{t+1}) + w_{t+1}^*(\sigma_{t+1}) \right] \quad (0 \le t < h).
\end{aligned}
$$

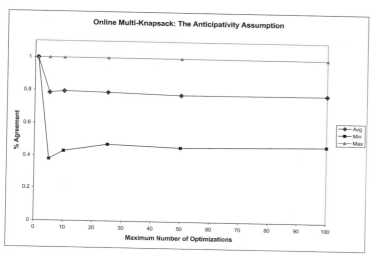

**Figure 13.3:** $\epsilon$-Anticipativity in Online Multiknapsacks.

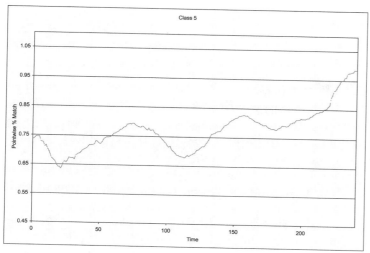

**Figure 13.4:** $\epsilon$-Anticipativity in Online Vehicle Routing.

ALGORITHM $\mathcal{E}^h(\sigma_t, i_t)$
1   **return** arg-max$(\sigma_{t+1} \in T(\sigma_t, i_t))$ $(R(\sigma_t, i_t, \sigma_{t+1}) + \text{EVALSTATE-}\mathcal{E}^h(t+1, \sigma_{t+1}))$;

FUNCTION EVALSTATE-$\mathcal{E}^h(t, \sigma_t)$
1   **if** $t = h$ **then**
2     **return** 0;
3   **else**
4     $e \leftarrow 0$;
5     **for** $k \leftarrow 1 \ldots m$ **do**
6       $i_t \leftarrow \text{SAMPLE}(t, t)$;
7       $e \leftarrow e + \max(\sigma_{t+1} \in T(\sigma_t, i_t))$ $(R(\sigma_t, i_t, \sigma_{t+1}) + \text{EVALSTATE-}\mathcal{E}^h(t+1, \sigma_{t+1}))$;
8     **return** $e/m$;

**Figure 13.5:** The Algorithm $\mathcal{E}^h$ for MCDPs.

Note also that the optimal decision $\pi^*(\sigma_t, i_t)$ at time $t$ is given by

$$\pi_t^*(\sigma_t, i_t) = \underset{\sigma_{t+1} \in T(\sigma_t, i_t)}{\text{arg-max}} \; (R(\sigma_t, \xi_t, \sigma_{t+1}) + w_{t+1}^*(\sigma_{t+1})).$$

More precisely, under similar assumptions on the variance, algorithm $\mathcal{E}^h$ produces a run

$$\mathcal{R} = \left\langle \sigma_0 \xrightarrow{i_0} \cdots \xrightarrow{i_{h-1}} \sigma_h \right\rangle$$

whose value $w(\mathcal{R})$ satisfies

$$|w_t^*(\sigma_0) - w(\mathcal{R})| \leq \epsilon$$

with a number $m$ of samples polynomial in $h$, $\max_{s,i} |T(s, i)|$, and $\frac{1}{\epsilon}$. Algorithm $\mathcal{E}^h$ is in fact close to the sparse sampling algorithm for MDCPs proposed by Kearns et al. [59] (see also [58, 76] for generalizations to POMDPs). The sparse sampling algorithm provides a similar result under various assumptions on the rewards and discounting for MDCPs with generative models. However, these algorithms run in $O((|I|m)^h)$ and are thus impractical both theoretically and experimentally even on small instances (see, for instance, [28] for an experimental evaluation). In fact, $\mathcal{E}^h$ exhaustively explores the search space of the underlying optimization problem, adding an extra level at each node to handle the sampling.

MCDPs, however, have several fundamental advantages over MDCPs and hence offer some opportunities when time constraints are not too severe. In particular, MCDPs clearly separate the optimization and the uncertainty models, which are both available separately. As a consequence, the anticipatory relaxation is always available to provide an optimistic evaluation of $w_t^*(\sigma_t)$. It is thus natural to propose the algorithm $\mathcal{E}^k$ $(k \geq 1)$ that selects the decision at time $t$ by exploring a

ALGORITHM $\mathcal{E}^k(\sigma_t, i_t)$
1  **return** arg-max$(\sigma_{t+1} \in T(\sigma_t, i_t))$ $(R(\sigma_t, i_t, \sigma_{t+1}) + \text{EVALSTATE-}\mathcal{E}^k(t+1, \sigma_{t+1}))$;

FUNCTION EVALSTATE-$\mathcal{E}^k(t, \sigma_t)$
1  **if** $t = h$ **then**
2     **return** 0;
3  **else**
4     **if** $k = 1$ **then**
5        **for** $j \leftarrow 1 \ldots m$ **do**
6           $i_{t..h-1} \leftarrow \text{SAMPLE}(t, h-1)$;
7           $e \leftarrow e + \mathcal{O}(\sigma_t, i_{t..h-1})$;
8        **return** $e/m$;
9     **else**
10       $e \leftarrow 0$;
11       **for** $i \leftarrow 1 \ldots m$ **do**
12          $i_t \leftarrow \text{SAMPLE}(t, t)$;
13          $e \leftarrow e + \max(\sigma_{t+1} \in T(\sigma_t, i_t))$ $(R(\sigma_t, i_t, \sigma_{t+1}) + \text{EVALSTATE-}\mathcal{E}^{k-1}(t+1, \sigma_{t+1}))$;
14       **return** $e/m$;

**Figure 13.6:** The Algorithm $\mathcal{E}^k$ $(k \geq 1)$ for Online Decision Processes.

tree depth of at most $k$. The algorithm is depicted in figure 13.6. Its main difference is in lines 4 through 8 which, at depth $k$, sample the future to the horizon and use the optimization algorithm to provide an optimistic evaluation

$$\mathcal{O}(\sigma_t, i_{t+1..h-1})$$

of the evaluation

$$\max(\sigma_{t+1} \in T(\sigma_t, i_t))\ (R(\sigma_t, i_t, \sigma_{t+1}) + \text{EVALSTATE-}\mathcal{E}^{k-1}(t+1, \sigma_{t+1})).$$

More generally, the exploration of such a tree should itself be viewed as an optimization problem where a node at depth $d$ can be pruned away if its optimistic evaluation at depth $d$ is worse than the evaluation of an existing leaf at depth $k$. In addition, the anticipatory relaxation can prune entire subtrees: if a first-level decision $s_1$ has been fully evaluated and the anticipatory relaxation of another decision $s_2$ is worse than the $s_1$'s evaluation, it is useless to evaluate the subtree rooted at $s_2$. The resulting algorithms will have similarities with real-time dynamic programming [3], the RTBSS algorithm [84] and, at the limit, with the sampling algorithm from [58].

## 13.8  The General Approximation Theorem for MCDPs

For completeness, we describe the results on algorithm $\mathcal{E}^h$. In the following, we use

$$T_* = \max_{\substack{s \in S \\ i \in I}} |T(s, i)|$$

and

$$w_* = \max_{\substack{t \in 0..h-1 \\ s_t \in S_t}} w_t(s_t).$$

The first step of proof shows that the value $w_t^*(\sigma_t)$ can be approximated closely with a polynomial number of samples $m$. This part of the proof follows the results by Kearns et al., but for a finite horizon and no discount. Algorithm $\mathcal{E}^h$ can be viewed as the following set of recursive equations

$$w_h(\sigma_h) = 0;$$
$$w_t(\sigma_t) = \frac{1}{m} \sum_{k=1}^{m} \max_{\sigma_{t+1} \in T(\sigma_t, i_t^k)} (R(\sigma_t, i_t^k, \sigma_{t+1}) + w_{t+1}(\sigma_{t+1})) \quad (t < h).$$

It is also interesting to define

$$\tilde{w}_t(\sigma_t) = \frac{1}{m} \sum_{k=1}^{m} \max_{\sigma_{t+1} \in T(\sigma_t, i_t^k)} (R(\sigma_t, i_t^k, \sigma_{t+1}) + w_{t+1}^*(\sigma_{t+1})) \quad (t < h).$$

Intuitively, $\tilde{w}_t(\sigma_t)$ is computed by sampling the future and using the optimal value $w_{t+1}^*(\sigma_{t+1})$ for evaluating subsequent steps. The $\tilde{w}_t$ values are useful in capturing the sampling errors.

LEMMA 13.7 (SAMPLING ERROR) For any $t \in H$ and any state $\sigma_t$, the relation

$$|w_t^*(\sigma_t) - \tilde{w}_t(\sigma_t)| \leq \lambda$$

holds with probability at least $1 - e^{-\frac{m\lambda^2}{w_*^2}}$.

PROOF:  By Hoeffding's inequality [53] and by definition of $w_t^*$ and $\tilde{w}_t$.  □

The next lemma specifies how the estimation error accumulates.

LEMMA 13.8 (ESTIMATION ERROR) For any $t \in H$ and any state $\sigma_t$, the relation

$$|w_t^*(\sigma_t) - w_t(\sigma_t)| \leq n\,\lambda$$

with probability at least

$$1 - (T_* m)^n e^{-\frac{m\lambda^2}{\sigma_*^2}}$$

where $n = h - t - 1$.

PROOF: By induction on $n$. The base case ($n = 0$) (and thus $t = h - 1$) holds by lemma 13.7. Consider now the inductive case:

$$|w_t^*(\sigma_t) - w_t(\sigma_t)| \leq |w_t^*(\sigma_t) - \tilde{w}_t(\sigma_t)| + |\tilde{w}_t(\sigma_t) - w_t(\sigma_t)|.$$

By lemma 13.7, $|w_t^*(\sigma_{t-1}) - \tilde{w}_t(\sigma_t)| \leq \lambda$ with probability at least $1 - e^{-\frac{m\lambda^2}{w_*^2}}$. Moreover,

$$|\tilde{w}_t(\sigma_t) - w_t(\sigma_t)|$$

is smaller or equal to

$$\frac{1}{m} \sum_{k=1}^{m} |\max_{\sigma_{t+1} \in T(\sigma_t, i_t^k)} (R(\sigma_t, i_t^k, \sigma_{t+1}) + w_{t+1}^*(\sigma_{t+1})) - \max_{\sigma_{t+1} \in T(\sigma_t, i_t^k)} (R(\sigma_t, i_t^k, \sigma_{t+1}) + w_{t+1}(\sigma_{t+1}))|.$$

By induction,

$$|w_{t+1}^*(\sigma_{t+1}) - w_{t+1}(\sigma_{t+1})| \leq (n-1)\lambda$$

with probability at least $1 - (T_* m)^{n-1} e^{-\frac{m\lambda^2}{w_*^2}}$. Since there are at most $T_*$ elements in $T(\sigma_t, i_t^k)$ and hence $mT_*$ such conditions,

$$|\tilde{w}_t(\sigma_t) - w_t(\sigma_t)| \leq (n-1)\lambda$$

holds with probability $1 - (T_* m)(T_* m)^{n-1} e^{-\frac{m\lambda^2}{w_*^2}}$. As a consequence,

$$|w_t^*(\sigma_t) - w_t(\sigma_t)| \leq n\,\lambda$$

with probability at least $1 - (T_* m)^n e^{-\frac{m\lambda^2}{w_*^2}}$. $\qquad \square$

Lemma 13.8 can be used to bound the error of $w_t(\sigma_t)$ with high probability. Indeed by choosing

$$m = \frac{w_*^2}{\lambda^2}\left(2h \log \frac{bh w_*^2}{\lambda^2} + \log \frac{1}{\delta}\right)$$

where $\delta = \frac{\lambda}{w_*}$, it follows that the inequality

$$|w_t^*(\sigma_t) - w_t(\sigma_t)| \leq (h - t) \lambda$$

holds with probability as least $1 - \delta$. At this stage, it remains to estimate the errors accumulated by algorithm $\mathcal{E}^h$. We start by defining the concept of local w-error.

**Definition 13.12 (Local w-Error at Time $t$)** The *local w-error* of $s_{t+1}$ at time $t$ for $(s_t, i_t)$, denoted by $\nabla(s_t, i_t, s_{t+1})$, is defined as

$$\max_{s_{t+1}^* \in T(s_t, i_t)} (R(s_t, i_t, s_{t+1}^*) + w_{t+1}^*(s_t^*)) - (R(s_t, i_t, s_{t+1}) + w_{t+1}^*(s_t)).$$

Once again, the total error of the algorithm is the summation of the local w-errors. Consider the sequence of random variables representing the decisions taken by algorithm $\mathcal{E}^h$:

$$\begin{aligned}
\sigma_1 &= \mathcal{E}^h(\sigma_0, \xi_0) \\
\sigma_2 &= \mathcal{E}^h(\sigma_1, \xi_1) \\
&\cdots \\
\sigma_h &= \mathcal{E}^h(\sigma_{h-1}, \xi_{h-1})
\end{aligned}$$

and let $\mathcal{R}$ denote the run

$$\mathcal{R} = \langle \sigma_0 \xrightarrow{\xi_0} \cdots \xrightarrow{\xi_{h-1}} \sigma_h \rangle.$$

---

LEMMA 13.9 (SUMMATION OF LOCAL W-ERRORS)

$$\sum_{t=0}^{h-1} \underset{\substack{\sigma_t, \xi_t \\ \sigma_{t+1}}}{\mathbb{E}} [\nabla(\sigma_t, \xi_t, \sigma_{t+1})] = w_0^*(\sigma_0) - \underset{\mathcal{R}}{\mathbb{E}}[w(\mathcal{E}^h(\mathcal{R}))].$$

---

PROOF:  Observe that

$$\underset{\substack{\sigma_t, \xi_t \\ \sigma_{t+1}}}{\mathbb{E}} \nabla(\sigma_t, \xi_t, \sigma_{t+1})$$

can be rewritten as

$$\underset{\substack{\sigma_t, \xi_t \\ \sigma_{t+1}}}{\mathbb{E}} [\max_{\sigma_{t+1}^* \in T(\sigma_t, \xi_t)} (R(\sigma_t, \xi_t, \sigma_{t+1}^*) + w_{t+1}^*(\sigma_{t+1}^*)) - (R(\sigma_t, \xi_t, \sigma_{t+1}) + w_{t+1}^*(\sigma_{t+1}))],$$

which is equivalent to

$$
\mathop{\mathbb{E}}_{\sigma_t} \mathop{\mathbb{E}}_{\xi_t} \max_{\sigma_{t+1}^* \in T(\sigma_t, \xi_t)} (R(\sigma_t, \xi_t, \sigma_{t+1}^*) + w_{t+1}^*(\sigma_{t+1}^*)) - \mathop{\mathbb{E}}_{\substack{\sigma_t, \xi_t \\ \sigma_{t+1}}} (R(\sigma_t, \xi_t, \sigma_{t+1}) + w_{t+1}^*(\sigma_{t+1}))
$$

$$
= \mathop{\mathbb{E}}_{\sigma_t} w_t^*(\sigma_t^*) - \mathop{\mathbb{E}}_{\substack{\sigma_t, \xi_t \\ \sigma_{t+1}}} R(\sigma_t, \xi_t, \sigma_{t+1}) - \mathop{\mathbb{E}}_{\sigma_{t+1}} w_{t+1}^*(\sigma_{t+1}).
$$

As a result,

$$
\sum_{t=1}^{h} \mathop{\mathbb{E}}_{\substack{\sigma_t, \xi_t \\ \sigma_{t+1}}} \nabla(\sigma_t, \xi_t, \sigma_{t+1}) = w_0^*(\sigma_0) - \sum_{t=0}^{h-1} \mathop{\mathbb{E}}_{\substack{\sigma_t, \xi_t \\ \sigma_{t+1}}} R(\sigma_t, \xi_t, \sigma_{t+1}) - \mathop{\mathbb{E}}_{\sigma_h} w_h^*(\sigma_h).
$$

$\square$

The last lemma bounds the local w-errors.

LEMMA 13.10 (LOCAL W-ERROR) Let $s_t \in S_t$, $i_t \in I$, $n = h - t - 1$ and $\overline{s}_{t+1}$ be the random variable $\mathcal{E}^h(s_t, i_t)$. The expected local w-error at step $t$ of algorithm $\mathcal{E}^h$ satisfies

$$
\mathop{\mathbb{E}}_{\overline{s}_{t+1}} [\nabla(s_t, i_t, \overline{s}_{t+1})] \le 2(n\lambda + \lambda).
$$

PROOF:  Let

$$
s_{t+1}^* = \mathop{\text{arg-max}}_{\sigma_{t+1} \in T(s_t, i_t)} (R(s_t, i_t, \sigma_{t+1}) + w_{t+1}^*(\sigma_{t+1})).
$$

We are interested in bounding

$$
R(s_t, i_t, s_{t+1}^*) + w_{t+1}^*(s_{t+1}^*) - (R(s_t, i_t, \overline{s}_{t+1}) + w_{t+1}^*(\overline{s}_{t+1})).
$$

Since $\mathcal{E}^h$ selects $\overline{s}_{t+1}$ over $s_{t+1}^*$, it follows that

$$
R(s_t, i_t, s_{t+1}^*) + w_{t+1}(s_{t+1}^*) \le R(s_t, i_t, \overline{s}_{t+1}) + w_{t+1}(\overline{s}_{t+1})
$$

and thus

$$
R(s_t, i_t, s_{t+1}^*) - R(s_t, i_t, \overline{s}_{t+1}) \le w_{t+1}(\overline{s}_{t+1}) - w_{t+1}(s_{t+1}^*).
$$

Hence, $R(s_t, i_t, s_{t+1}^*) + w_{t+1}^*(s_{t+1}^*) - (R(s_t, i_t, \overline{s}_{t+1}) + w_{t+1}^*(\overline{s}_{t+1}))$ is bounded by

$$
\begin{aligned}
& w_{t+1}(\overline{s}_{t+1}) - w_{t+1}^*(\overline{s}_{t+1}) + w_{t+1}^*(s_{t+1}^*) - w_{t+1}(s_{t+1}^*) \\
\le\ & |w_{t+1}(\overline{s}_{t+1}) - w_{t+1}^*(\overline{s}_{t+1})| + |w_{t+1}^*(s_{t+1}^*) - w_{t+1}(s_{t+1}^*)|.
\end{aligned}
$$

By lemma 13.8, the expected error of a state $\bar{s}_{t+1}$ is bounded by

$$2((1-\delta)n\lambda + \delta w_*) \le 2(n\lambda + \lambda).$$

□

THEOREM 13.2 (QUALITY OF $\mathcal{E}^h$) Let $\epsilon > 0$. Algorithm $\mathcal{E}^h$ returns a state $\sigma_h$ satisfying

$$|w_1^*(\sigma_\perp) - w(\sigma_h)| \le \epsilon$$

using a number of samples that is polynomial in $h$, $\max_{s,i} T(s,i)$, and $\frac{1}{\epsilon}$.

PROOF:  By lemmas 13.9 and 13.10, the total expected loss of $\mathcal{E}^h$ with respect to $w_0^*(\sigma_\perp)$ is bounded by

$$2h(h\lambda + \lambda) = 2(h^2 + h)\lambda.$$

The result follows by choosing $\lambda = \frac{\epsilon}{4h^2}$.

□

## 13.9  Notes and Further Reading

A stochastic-programming formulation of the main result in this chapter is given in [78] The paper also discusses the relationship between the anticipatory gap and the expected value of perfect information, a traditional concept in stochastic programming. Moreover, the paper analyzes the anticipatory gap on a simplified version of packet scheduling, providing a theoretical justification for small anticipatory gap.

# References

[1] A. Anagnostopoulos, R. Bent, E. Upfal, and P. Van Hentenryck. A Simple and Deterministic Cfompetitive Algorithm for Online Facility Location. *Information and Computation*, 194(2):175–202, November 2004.

[2] N. Ascheuer, S. Krumke, and J. Rambau. Online Dial-A-Ride Problems: Minimizing the Completion Time. In *Proceedings of the 17th Annual Symposium on Theoretical Aspects of Computer Science (STACS'00)*, pp. 639–650. Springer Verlag, 2000.

[3] A. Barto, S. Bradtke, and S. Singh. Learning to Act Using Real-Time Dynamic Programming. *Artificial Intelligence*, 72(1–2):81–138, 1995.

[4] E. Baum. An Inequality and Associated Maximization Technique in Statistical Estimation for Probabilistic Functions of a Markov Process. *Inequalities*, 3:1–8, 1972.

[5] J. E. Beasley. http://people.brunel.ac.uk/~mastjjb/jeb/or/sp.html.

[6] T. Benoist, E. Bourreau, Y. Caseau, and B. Rottembourg. Towards Stochastic Constraint Programming: A Study of Online Multi-choice Knapsack with Deadlines. In *Proceedings of the Seventh International Conference on Principles and Practice of Constraint Programming (CP'01)*, pp. 61–76. Springer Verlag, 2001.

[7] R. Bent, I. Katriel, and P. Van Hentenryck. Sub-Optimality Approximation. In *Eleventh International Conference on Principles and Practice of Constraint Programming (CP'05)*, pp. 122–136. Springer Verlag, 2005.

[8] R. Bent and P. Van Hentenryck. Dynamic Vehicle Routing with Stochastic requests. In *International Joint Conference on Artificial Intelligence (IJCAI'03)*. AAAI Press, 2003.

[9] R. Bent and P. Van Hentenryck. A Two-Stage Hybrid Local Search for the Vehicle Routing Problem with Time Windows. *Transportation Science*, 8(4):515–530, 2004.

[10] R. Bent and P. Van Hentenryck. Online Stochastic and Robust Optimization. In *Proceeding of the 9th Asian Computing Science Conference (ASIAN'04)*. Springer Verlag, December 2004.

[11] R. Bent and P. Van Hentenryck. Regrets Only. Online Stochastic Optimization under Time Constraints. In *Proceedings of the 19th National Conference on Artificial Intelligence (AAAI'04)*. AAAI Press, July 2004.

[12] R. Bent and P. Van Hentenryck. Scenario-Based Planning for Partially Dynamic Vehicle Routing Problems with Stochastic Customers. *Operations Research*, 52(6), 2004.

[13] R. Bent and P. Van Hentenryck. The Value of Consensus in Online Stochastic Scheduling. In *Proceedings of the 14th International Conference on Automated Planning and Scheduling (ICAPS 2004)*. AAAI Press, 2004.

[14] R. Bent and P. Van Hentenryck. Online Stochastic Optimization without Distributions. In *Proceedings of the 15th International Conference on Automated Planning and Scheduling (ICAPS 2005)*. AAAI Press, 2005.

[15] R. Bent and P. Van Hentenryck. A Two-Stage Hybrid Algorithm for Pickup and Delivery Vehicle Routing Problems with Time Windows. *Computers and Operations Research (Special Issue on Applications in Combinatorial Optimization)*, pp. 875–893, 2006.

[16] D. Bertsekas and J. Tsitsiklis. *Neuro-Dynamic Programming*. Athena Scientific, Belmont, MA, 1996.

[17] D. Bertsimas. A Vehicle Routing Problem with Stochastic Demand. *Operations Research*, 40:574–585, 1992.

[18] D. Bertsimas and G. Van Ryzin. A Stochastic and Dynamic Vehicle Routing Problem in the Euclidean Plane. *Operations Research*, 39:601–615, 1991.

[19] D. Bertsimas and G. Van Ryzin. Stochastic and Dynamic Vehicle Routing in the Euclidean Plane with Multiple Capacitated Vehicles. *Operations Research*, 41:60–76, 1993.

[20] J. R. Birge and F. Louveaux. *Introduction to Stochastic Programming*. Springer, New York, NY, 1997.

[21] C. Blair and R. Jeroslow. On the Value Function of an Integer Program. *Mathematical Programming*, 23, 1979.

[22] A. Borodin and R. El-Yaniv. *Online Computation and Competitive Analysis*. Cambridge University Press, New York, NY, 1998.

[23] A. Borodin, J. Kleinberg, P. Raghawan, M. Sudan, and D. Williamson. Adversarial Queuing Theory. *JACM*, 48(1):13–38, 2001.

[24] H. Burckert, K. Fischer, and G. Vierke. Holonic Transport Scheduling with Teletruck. *Applied Artificial Intelligence*, 14:697–725, 2000.

[25] A. Campbell and M. Savelsbergh. Decision Support for Consumer Direct Grocery Initiatives. *Report TLI-02-09, Georgia Institute of Technology*, 2002.

[26] A. Campbell and M. Savelsbergh. Incentive Schemes for Attended Home Delivery Services. *Technical Report, Georgia Institute of Technology*, 2003.

[27] C. Caroe and G. Schultz. Dual Decomposition in Stochastic Integer Programming. *Operations Research Letters*, 24(1–2):37–45, 1999.

[28] H. Chang, R. Givan, and E. Chong. Online Scheduling Via Sampling. *Artificial Intelligence Planning and Scheduling (AIPS'00)*, pp. 62–71, 2000.

[29] E. Charniak. *Statistical Language Learning*. The MIT Press, Cambridge, MA, 1993.

[30] C. Chatfield. *The Analysis of Time Series: An Introduction*. Chapman and Hall, Norwell, MA, 2004.

[31] J. Choi, M. Realff, and J. Lee. Dynamic Programming in a Heuristically Confined State Space: A Stochastic Resource-Constrained Project Scheduling Application. *Computers and Chemical Engineering*, 28(6–7):1039–1058, 2004.

[32] J. Csirik, D. Johnson, C. Kenyon, J. Orlin, P. Shor, and R. Weber. On the Sum-of-Squares Algorithm for Bin Packing. In *Proceedings of the 32nd Annual ACM Symposium on Theory of Computing (STOCS'2000)*, pp. 208–217. ACM Press, 2000.

[33] M. Daskin and L. Dean. Location of Health Care Facilities. In M. Brendeau, F. Sainfort, and W. Perskalla, editors, *Operations Research and Health Care*, pp. 43–76. Kluwer Academic Publishers, Norwell, MA, 2004.

[34] D. de Farias and B. Van Roy. Approximate Linear Programming for Average-Cost Dynamic Programming. In *Advances in Neural Information Processing Systems 15 (NIPS 2002)*, pp. 1587–1594. MIT Press, 2002.

[35] B. Dean, M. Goemans, and J. Vondrák. Approximating the Stochastic Knapsack Problem: The Benefit of Adaptivity. In *Proceedings of the 45th Annual IEEE Symposium on Foundations of Computer Science*, pp. 208–217, Rome, Italy, 2004. ACM Press.

[36] A. Fiat and G. Woeginger. *Online Algorithms: The State of the Art.* 1998.

[37] M. Fisher, K. Jornsten, and O. Madsen. Vehicle Routing with Time Windows: Two Optimization Algorithms. *Operations Research*, 45:488–492, 1997.

[38] A. Flaxman, A. Frieze, and M. Krivelevich. On the Random 2-Stage Minimum Spanning Tree. In *ACM-SIAM Symposium on Discrete Algorithms (SODA-2005)*. ACM Press, 2005.

[39] M. Gendreau, F. Guertin, J. Y. Potvin, and R. Seguin. Neighborhood Search Heuristics for a Dynamic Vehicle Dispatching Problem with Pick-ups and Deliveries. *Technical Rep. CRT-98-10, Centre de Recherche sur les Transport, Université de Montreal*, 1998.

[40] M. Gendreau, F. Guertin, J. Y. Potvin, and E. Taillard. Parallel Tabu Search for Real-Time Vehicle Routing and Dispatching. *Transportation Science*, 33 (4):381–390, 1999.

[41] M. Gendreau, G. Laporte, and R. Seguin. An Exact Algorithm for the Vehicle Routing Problem with Stochastic Demands and Customers. *Transportation Science*, 29:143–155, 1995.

[42] M. Gendreau, G. Laporte, and R. Seguin. A Tabu Search Heuristic for the Vehicle Routing Problem with Stochastic Demands and Customers. *Operations Research*, 44:469–477, 1996.

[43] M. Gendreau, G. Laporte, and R. Seguin. Stochastic Vehicle Routing. *European Journal of Operational Research*, 88:3–12, 1996.

[44] M. Gendreau, G. Laporte, and F. Semet. A Dynamic Model and Parallel Tabu Search Heuristic for Real-Time Ambulance Relocation. *Parallel Computing*, 27:1641–1653, 2001.

[45] M. Goemans and J. Vondrák. Stochastic Covering and Adaptivity. In J. R. Correa, A. Hevia, and M. A. Kiwi, editors, *LATIN*, vol. 3887 of *Lecture Notes in Computer Science*, pp. 532–543. Springer, 2006.

[46] L. Green. Capacity Planning and Management in Hospitals. In M. Brendeau, F. Sainfort, and W. Perskalla, editors, *Operations Research and Health Care*, pp. 15–42. Kluwer Academic Publishers, Norwell, MA, 2004.

[47] Y. Guan, S. Ahmed, A. Miller, and G. Nemhauser. On Formulations of the Stochastic Uncapacitated Lot-Sizing Problem. *Operations Research Letters*, 34(3):241–250, May 2006.

[48] Y. Guan, S. Ahmed, G. Nemhauser, and A. Miller. A Branch-and-Cut Algorithm for the Stochastic Uncapacitated Lot-Sizing Problem. *Mathematical Programming*, 105:55–84, 2006.

[49] P. Hansen and N. Mladenovic. An Introduction to Variable Neighborhood Search. In S. Voss, S. Martello, I. H. Osman, and C. Roucairol, editors, *Meta-Heuristics: Advances and Trends in Local Search Paradigms for Optimization*, pp. 433–458. Kluwer Academic Publishers, Norwell, MA, 1998.

[50] W. D. Harvey and M. L. Ginsberg. Limited Discrepancy Search. In *Proceedings of the 14th International Joint Conference on Artificial Intelligence*. Morgan Kaufmann, August 1995.

[51] D. Hauptmeier, S. Krumke, and J. Rambau. The Online Dial-A-Ride Problem Under Reasonable Load. In *Proceedings of the 4th Italian Conference on Algorithms and Complexity (CIAC-2000), Rome, Italy*, pp. 125–136. Springer Verlag, 2000.

[52] J. L. Higle and S. Sen. *Stochastic Decomposition*. Kluwer Academic Publishers, 1996.

[53] W. Hoeffding. Probability Inequalities for Sums of Bounded Random Variables. *Journal of the American Statistical Association*, 58(301):13–30, 1963.

[54] S. Ichoua, M. Gendreau, and J.Y. Potvin. Diversion Issues in Real-Time Vehicle Dispatching. *Transporation Science*, 34:426–438, 2000.

[55] L. Kaelbling, M. Littman, and A. Cassandra. Planning and Acting in Partially Observable Stochastic Domains. *Artificial Intelligence*, 101 (1–2):99–134, 1998.

[56] P. Kall and S. W. Wallace. *Stochastic Programming*. John Wiley and Sons, Chichester, England, 1994.

[57] A. Karlin, M. Manasse, L. Rudolph, and D. Sleator. Competitive Snoopy Caching. *Algorithmica*, 3:79–119, 1988.

[58] M. Kearns, Y. Mansour, and A. Ng. A Sparse Sampling Alogorithm for Near-Optimal Planning in Large Markov Decision Processes. *Proceedings of the International Joint Conference on Artificial Intelligence (IJCAI'99), Stockholm, Sweden*, pp. 1324–1231, 1999.

[59] M. Kearns, Y. Mansour, and A. Ng. A Sparse Sampling Algorithm for Near-Optimal Planning in Large Markov Decision Processes. *Machine Learning*, 49(2-3):193–208, 2002.

[60] A. Kesselman, Z. Lotker, Y. Mansour, B. Patt-Shamir, B. Schieber, and M. Sviridenko. Buffer Overflow Management in QoQ Switches. *SIAM Journal on Computing*, 33:563–583, 2004.

[61] P. Kilby, P. Prosser, and P. Shaw. Dynamic VRPs: A Study of Scenarios. *Technical Report APES-06-1998*, 1998.

[62] G. Kindervater and M. Savelsbergh. Vehicle Routing: Handling Edge Exchanges. In E. Aarts and J.K. Lenstra, editors, *Local Search in Combinatorial Optimization*, chapter 10, pp. 337–360. John Wiley and Sons, 1997.

[63] A. Kleywegt, V. Nori, and M. Savelsbergh. Dynamic Programming Approximations for a Stochastic Inventory Routing Problem. *Transportation Science*, 38:42–70, 2004.

[64] A. Kleywegt and A. Shapiro. The Sample Average Approximation Method for Stochastic Discrete Optimization. *SIAM Journal on Optimization*, 15(1–2):1–30, January 1999.

[65] N. Kohl, J. Desrosiers, O. Madsen, M. Solomon, and F. Soumis. 2-Path Cuts for the Vehicle Routing Problem with Time Windows. *Transportation Science*, 33:101–116, 1999.

[66] E. Koutsoupias and C. H. Papadimitriou. Beyond Competitive Analysis. *SIAM Journal on Computing*, 30(1):300–317, 2000.

[67] S. Krumke, J. Rambau, and L. Torres. Real-Time Dispatching of Guided and Unguided Automobile Service Units with Soft Time Windows. In *Proceedings of 10th Annual European Symposium on Algorithms (ESA-2002)*, pp. 637–648. Springer Verlag, 2002.

[68] G. Laporte, F. Louveaux, and H. Mercure. The Vehicle Routing Problem with Stochastic Travel Times. *Transportation Science*, 26:161–170, 1992.

[69] A. Larsen. The Dynamic Vehicle Routing Problem. *Ph.D. Thesis, Technical University of Denmark*, 2000.

[70] A. Larsen, O. Madsen, and M. Solomon. Partially Dynamic Vehicle Routing Models and Algorithms. *Journal of Operational Research Society*, 53:637–646, 2002.

[71] A. Larsen, O. Madsen, and M. Solomon. The A-Priori Dynamic Traveling Salesman Problem with Time Windows. *Transportation Science*, 38:459–472, 2004.

[72] C. Lefevre. Optimal Control of a Birth and Death Epidemic Process. *Operations Research*, 29:971–982, 1981.

[73] R. Levi, M. Pal, R. Roundy, and D. Shmoys. Approximation Algorithms for Stochastic Inventory Control Models. In *Proceedings of the 11th MPS Conference on Integer Programming and Combinatorial Optimization (IPCO'05)*, pp. 306–32. Springer Verlag, June 2005.

[74] J. Linderoth, A. Shapiro, and S. Wright. The Empirical Behavior of Sampling Methods for Stochastic Programming. *Annals of Operations Research*, 2006. (to appear).

[75] C. Lund and M. Yannakakis. On the Hardness of Approximating Minimization Problems. *Journal of the ACM*, 41:60–981, 1994.

[76] D. McAllester and S. Singh. Approximate Planning for Factored POMDPs Using Belief State Simplification. In *Proceedings of the 15th Conference on Uncertainty in Artificial Intelligence (UAI-99)*, pp. 409–416, San Francisco, CA, AAAI Press, August 1999.

[77] Luc Mercier and P. Van Hentenryck. Online Reservation Systems Revisited. Technical report, Brown University, September 2006.

[78] Luc Mercier and P. Van Hentenryck. Performance Analysis of Online Anticipatory Algorithms for Large Multistage Stochastic Integer Programs. Technical report, Brown University, September 2006.

[79] A. Meyerson. Online Facility Location. In *42nd IEEE Symposium on Foundations of Computer Science (FOCS-01)*, pp. 14–17, Las Vegas, Nevada, ACM Press, October 2001.

[80] M. Mitrovic-Minic, R. Krishnamurti, and G. Laporte. Double-Horizon–Based Heuristics for the Dynamic Pickup and Delivery Problems with Time Windows. *Transportation Research Record Part B*, 38:669–685, 2004.

[81] M. Mitrovic-Minic and G. Laporte. Waiting Strategies for the Dynamic Pickup and Delivery Problem with Time Windows. *Transportation Research Record Part B*, 38:635–655, 2004.

[82] R. H. Moehring, F. J. Radermacher, and G Weiss. Stochastic Scheduling Problems I: General Strategies. *ZOR - Zeitschrift fuer Operations Research*, 28:193–260, 2004.

[83] R. H. Moehring, A. Schulz, and M. Uetz. Approximation in Stochastic Scheduling: The Power of LP-Based Priority Policies. *Journal of the ACM*, 46(6):924–942, 1999.

[84] D. Nikovski and M. Branch. Marginalizing Out Future Passengers in Group Elevator Control. In *Proceedings of the 19th Conference in Uncertainty in Artificial Intelligence (UAI'03), Acapulco, Mexico*. Morgan Kaufmann, 2003.

[85] A. Novaes and O. Graciolli. Designing Multiple-Vehicle Delivery Tours in a Grid-Cell Format. *European Journal of Operational Research*, 119:613–634, 1999.

[86] S. Paquet, L. Tobin, and B. Chaib-draa. Real-Time Decision Making for Large POMDPs. In *18th Canadian Conference on Artificial Intelligence (AI'05)*. Springer Verlag, 2005.

[87] W. B. Powell and B. van Roy. Approximate Dynamic Programming for High-Dimensional Dynamic Resource Allocation Problems. In *Handbook of Learning and Approximate Dynamic Programming*, pp. 261–279, 2004.

[88] H. Psaraftis. Dynamic Vehicle Routing: Status and Prospects. *Annals of Operations Research*, 61:143–164, 1995.

[89] M. Puterman. *Markov Decision Processes*. John Wiley and Sons, New York, 1994.

[90] B. Rexing, C. Barnhart, T. Kniker, A. Jarrah, and N. Krishnamurthy. Airline Fleet Assignment with Time Windows. *Transportation Science*, 34(1):1–20, 2000.

[91] W. Romisch and R. Schultz. Stability of Solutions for Stochastic Programs with Complete Recourse. *Mathematics of Operations Research*, 18:590–609, 1993.

[92] J. Rosenberger, E. Johnson, and G. Nemhauser. Rerouting Aircraft for Airline Recovery. *Transportation Science*, 37(4):408–421, November 2003.

[93] J. Rosenberger, E. Johnson, and G. Nemhauser. A Robust Fleet Assignment Model with Hub Isolation and Short Cycles. *Transportation Science*, 38(3):357–368, August 2004.

[94] S. Ross. *A First Course in Probability*. Prentice Hall, Englewood Cliffs, New Jersey, 1997.

[95] M. Savelsbergh and M. Sol. DRIVE: Dynamic Routing of Independent Vehicles. *Operations Research*, 46:474–490, 1998.

[96] R. Schultz. Continuity Properties of Expectation Functions in Stochastic Integer Programming. *Mathematics of Operations Research*, 18(3):578–589, 1993.

[97] R. Schultz. On Structure and Stability in Stochastic Programs with Random Technology Matrix and Complete Integer Recourse. *Mathematical Programming*, 70(1):73–89, 1995.

[98] R. Schultz, L. Stougie, and M. H. van der Vlerk. Two-Stage Stochastic Integer Programming: A Survey. *Statistica Neerlandica. Journal of the Netherlands Society for Statistics and Operations Research*, 50(3):404–416, 1996.

[99] R. Schultz, L. Stougie, and M. H. van der Vlerk. Solving Stochastic Programs with Integer Recourse by Enumeration: A Framework Using Gröbner Basis Reductions. *Mathematical Programming*, 83:229–252, 1998.

[100] N. Secomandi. Comparing Neuro-Dynamic Programming Algorithms for the Vehicle Routing Problem with Stochastic Demands. *Computers and Operations Research*, 27:1201–1225, 2000.

[101] N. Secomandi. A Rollout Policy for the Vehicle Routing Problem with Stochastic Demands. *Operations Research*, 49:796–802, 2001.

[102] A. Shapiro. On Complexity of Multistage Stochastic Programs. *Operations Research Letters*, 34(1):1–8, January 2006.

[103] A. Shapiro and T. Homem-de Mello. A Simulation-Based Approach to Two-Stage Stochastic Programming with Recourse. *Mathematical Programming*, 81(3, Ser. A):301–325, 1998.

[104] P. Shaw. Using Constraint Programming and Local Search Methods to Solve Vehicle Routing Problems. In *Proceedings of Fourth International Conference on the Principles and Practice of Constraint Programming (CP'98)*, pp. 417–431. Springer Verlag, October 1998.

[105] D. Shmoys and C. Swamy. Stochastic Optimization Is (Almost) as Easy as Deterministic Optimization. In *Proceedings of the 45th Symposium on Foundations of Computer Science (FOCS-2004)*, pp. 228–237. IEEE Computer Society, October 2004.

[106] M. Skutella and M. Uetz. Stochastic Machine Scheduling with Precedence Constraints. *SIAM Journal on Computing*, 34(4):788–802, 2005.

[107] M. Solomon. Algorithms for the Vehicle Routing and Scheduling Problems with Time Window Constraints. *Operations Research*, 35 (2):254–265, 1987.

[108] L. Stougie. *Design and Analysis of Methods for Stochastic Integer Programming*. PhD thesis, University of Amsterdam, 1985.

[109] L. Stougie and M. H. van der Vlerk. Stochastic Integer Programming. In *Annotated Bibliographies in Combinatorial Optimization*, M. Dell'Amico et al. (editor), John Wileyi and Sons, New York, pp. 127–141, 1997.

[110] C. Swamy and D. Shmoys. Sampling-based Approximation Algorithms for Multistage Stochastic Optimization. In *Proceedings of 46th Annual IEEE Symposium on Foundations of Computer Science (FOCS 2005)*, pp. 357–366. IEEE Computer Society, October 2005.

[111] M. Swihart and J. Papastavrou. A Stochastic and Dynamic Model for the Single-Vehicle Pickup and Delivery Problem. *European Journal of Operational Research*, 114:447–464, 1999.

[112] M. H. van der Vlerk. *Stochastic Programming with Integer Recourse*. PhD thesis, University of Groningen, The Netherlands, 1995.

[113] J. van Hemert and J. La Poutré. Dynamic Routing with Fruitful Regions: Models and Evolutionary Computation. In *Parallel Problem Solving from Nature VIII*, pp. 690–699. Springer Verlag, 2004.

[114] P. Van Hentenryck, R. Bent, and E. Upfal. Online Stochastic Optimization under Time Constraints. *Annals of Operations Research (Special Issue on Stochastic Programming)*, 2005. (Submitted).

[115] P. Van Hentenryck, R. Bent, and Y. Vergados. Online Reservation Systems. In *Proceedings of the Third International Conference on the Integration of AI and OR Techniques in Constraint Programming for Combinatorial Optimization Problems (CP-AI-OR'06)*. Springer Verlag, 2006.

[116] F. Villareal and R. Bulfin. Scheduling a Single Machine to Minimize the Weighted Number of Tardy Jobs. *IEEE Transactions*, pp. 337–343, 1983.

[117]  W. Yang, K. Mathur, and R. Ballou. Stochastic Vehicle Routing Problem with Restocking. *Transportation Science*, 34:99–112, 2000.

[118]  K. Zhu and K. Ong. A Reactive Method for Real Time Dynamic Vehicle Routing Problem. In *12th IEEE International Conference on Tools with Artificial Intelligence (ICTAI)*, pp. 176–181. IEEE Computer Society, 2000.

# Index